# Water Operator Certification Exam Prep

By John Giorgi
Reviewed by the Association of Boards of Certification

# Water Operator Certification Exam Prep



**Disclaimer**

Although this study guide has been extensively reviewed for accuracy, there may be an occasion to dispute an answer, either factually or in the interpretation of the question. Both AWWA and ABC have made every effort to correct or eliminate any questions that may be confusing or ambiguous. If you do find a question that you feel is confusing or incorrect, please contact the AWWA Publishing Group at books@awwa.org.

Additionally, it is important to understand the purpose of this study guide. It does not guarantee certification. It is intended to provide an operator, or aspiring operator, with an understanding of the types of questions he or she will be presented with on a certification exam and the areas of knowledge that will be covered. AWWA highly recommends that readers study and understand the reference material from which the questions are drawn.

Director of Publishing: Zsolt Silberer
Senior Manager, Editorial: Kimberly J. Retzlaff
Senior Manager, Production: Daniel Berger
Managing Editor, Book Products: Melissa Valentine
Product Manager: Tony Petrites
Manager of Publishing Operations: Megan McCarthy
Project Manager: Caitlin Mercier
Cover Design: Melanie Yamamoto

---

**Library of Congress Cataloging-in-Publication Data**

Has been applied for.

ISBN: 978-1-62576-264-1

Printed in the United States of America
American Water Works Association
6666 West Quincy Avenue
Denver, CO 80235-3098
**awwa.org**

# Contents



## Questions

## Answers

# About AWWA & ABC

## American Water Works Association

The American Water Works Association (AWWA) is an international, nonprofit, scientific, and educational society dedicated to providing total water solutions assuring the effective management of water. Founded in 1881, AWWA is the largest organization of water supply professionals in the world.

Our membership includes over 4,000 utilities that supply roughly 80 percent of the nation's drinking water and treat almost half of the nation's wastewater, and over 50,000 members that represent the full spectrum of the water community: public water and wastewater systems, environmental advocates, scientists, academicians, and others who hold a genuine interest in water, our most important resource.

AWWA also takes great pride in helping establish two preeminent organizations dedicated to safe water—the Water Research Foundation in 1966 and Water For People in 1991.

AWWA unites the diverse water community to advance public health and safety, the economy, and the environment.

What we do:

- Offer education to water professionals
- Advocate for safe and sustainable water
- Collect and share knowledge
- Create volunteering opportunities

## Association of Boards of Certification

The Association of Boards of Certification (ABC) was founded to advance the quality and integrity of environmental certification programs throughout the world. ABC believes in certification as a means of promoting public health and the environment, striving to give its members the tools to ensure operators carry the highest level of knowledge and skills. For more than 45 years, ABC has provided education and resources to nearly 100 certifying authorities representing more than 40 states, 10 Canadian provinces and territories, as well as several international and tribal programs.

ABC's Vision: Public health and the environment are protected through stewardship of water resources.

ABC's Mission: Competent water and wastewater professionals in a supportive policy environment

### *Certification Commission for Environmental Professionals*

The Certification Commission for Environmental Professionals (C2EP) is an independent governing body of the water environment industry and certification subject matter experts. It was established by ABC with the goal of enhancing the integrity and standardization of operator certification. C2EP has been instrumental in the creation of the Professional Operator (PO) certification and designation program—the industry's first and only internationally recognized professional designation.

By earning PO certification in water treatment, water distribution, wastewater treatment, or wastewater collection system operations, operators demonstrate their ability to meet international standards and the desire to grow as professionals. This, combined with mandatory adherence to the Professional Operator *Code of Conduct*, promotes public visibility, consumer confidence, and support of water professionals and the industry.

# Preface

This book was developed by the American Water Works Association (AWWA) with the cooperation of the Association of Boards of Certification (ABC). All of the questions are based on ABC's Need-to-Know Criteria for four grades of certification for water treatment operators and water distribution operators.

The book is split into four sections: *Introductory* sections for water treatment and water distribution cover material corresponding to ABC's grades 1 and 2, while *Advanced* sections for water treatment and water distribution cover material corresponding to ABC's grades 3 and 4.

All of the questions in this book are new to this edition and have been reviewed and vetted by volunteer experts at ABC. Math questions are worked out in steps, shown after each answer, to allow readers to check their work.

This guide is intended to give operators practice in answering questions that are similar in format and content to the questions that appear on certification examinations. While the questions found on an examination are not duplicated in this guide, similar types of questions are included; they cover relevant areas of study and allow operators to understand what areas they need to study to pass an examination. Operators who have difficultly answering any of the questions in this book should consult the reference source provided after the answer to the question.

## Study Tips

1. Use this book alongside other training materials and/or courses, not in place of them. The questions in this book are designed to test your knowledge and identify content areas in which you need to do further study. Testing yourself with the questions in this book is an excellent way to review your understanding of a topic.
2. If you are having trouble with a question or set of questions, consider consulting the specific reference mentioned after the answer for further study.
3. Do not memorize the questions and answers. The questions in this book will not be duplicated exactly in a certification examination. If you understand the context of a question, you should be able to answer any number of similar questions on the topic.
4. Check with your local certification agency (state or provincial) so you know what subjects and levels will be covered under your specific certification examination. There are significant regional and state/provincial variations in regulatory requirements, certification criteria, treatment plants, and distribution systems.

5. Give yourself as much time as you can. Studying works best when you have the time to review material several times.

## References and Resources

The following publications were consulted during the writing of this book. The publication most appropriate to each question is mentioned after each answer.

**Water System Operations (WSO) Series**

*WSO Water Distribution, Grades 1 & 2* (AWWA, 2016)

*WSO Water Distribution, Grades 3 & 4* (AWWA, 2016)

*WSO Water Treatment, Grade 1* (AWWA, 2016)

*WSO Water Treatment, Grade 2* (AWWA, 2016)

*WSO Water Treatment, Grades 3 & 4* (AWWA, 2016)

*Basic Microbiology for Drinking Water, Third Edition* (AWWA, 2014)

*Math for Distribution System Operators* (AWWA, 2007)

*Math for Water Treatment Operators* (AWWA, 2007)

*Pumps and Pumping, 10th Edition* (ACR Publications, 2010)

*Water Distribution Operator Training Handbook, Fourth Edition* (AWWA, 2013)

*Water Treatment Operator Training Handbook, Third Edition* (AWWA, 2013)

# Acknowledgments

The following individuals served as volunteer reviewers for the questions in this book. Their service and dedication to the water operator profession and the water industry are greatly appreciated.

Cheryl Capron, PO (Seattle Public Utilities)

Ari Copeland, PO (Black & Veatch)

Alan C. Cranford, PO (Murfreesboro Water Resources Department)

F. Dede deMarks, PhD, PO (DdM Professionals, LLC)

D. Kim Dyches, CPM (Utah Division of Drinking Water)

Trey Finch, PO (Birmingham Water Works Board)

Andrew Houlihan, CET (Halifax Water)

Darin M. LaFalam (City of Worcester, MA)

Shaun Livermore, PO (Poarch Band of Creek Indians Utilities Authority)

William J. Mabuce, PE, PO, ENV SP (West Des Moines Water Works)

Michael R. McClenathan (City of Central Point, OR)

Alison McGee, PO (Huntsville Utilities)

Timothy M. Meloveck, PO, CWP (Town of Carbondale, CO)

Martin Nutt (Arkansas Department of Health)

Daniel "Rick" Pfleiderer (Greenville Water)

Gary Sale, PO (Washington State Parks)

Brian Thorburn, PO, AscT (EOCP)

David Weihrauch (City of Oxford, OH)

Ben Wright, PO (City of Cayce, SC)

The authors would also like to thank the following individuals who helped improve the clarity and accessibility of the questions and answers in this book: Dana Butler, Kyla Jacobsen, Somlynn Rorie, and Carina Stanton.

This page intentionally blank.

# I. Water Treatment, Introductory Questions

## Treatment Process

1. What is the area of a circular clarifier in $ft^2$ if it has a diameter of 125 ft?

    a. 49,062 $ft^2$
    b. 12,226 $ft^2$
    c. 15,625 $ft^2$
    d. 24,531 $ft^2$

2. What is the surface loading rate for a sedimentation basin that is 125 ft by 68 ft if it is treating an instantaneous flow rate of 4.75 $ft^3/s$?

    a. 361 gpd/$ft^2$
    b. 48.3 gpd/$ft^2$
    c. 40,375 gpd/$ft^2$
    d. 0.0006 gpd/$ft^2$

3. What is the most common disinfection by product when adding chlorine to water containing natural and inorganic matter?

    a. Chloroform
    b. Bromodichloromethane
    c. Monochloroacetic acid
    d. Dichloroacetic acid

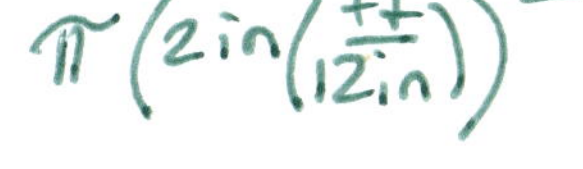

4. What is the velocity of flow in ft/s for a 4.0-in.-diameter pipe if it delivers 178 gpm?

    a. 14.2 ft/s
    b. 4.6 ft/s
    c. 0.032 ft/s
    d. 6.8 ft/s

5. What will most probably cause taste and odors, slime, and filter clogging?

    a. Decaying plant material
    b. Filamentous fungi
    c. Blue-green algae
    d. Some species of ciliated protozoa

6. Why can rapid sand filters handle higher filtration rates compared to slow sand filters?

   a. Finer sand
   b. Coarser sand
   c. Much higher water level above the filter's media, thus higher head pressure
   d. Media not as thick

7. What should the chemical feeder setting be in mL/min for a polymer solution if the desired dosage is 3.50 mg/L and the treatment plant is treating 13.7 mgd? The specific gravity of the polymer is 1.28.

   a. 97.3 mL/min of polymer
   b. 98.3 mL/min of polymer
   c. 98.5 mL/min of polymer
   d. 99.0 mL/min of polymer

8. Iron bacteria can interfere with __________ because the bacteria produce __________ that prevents the stabilization chemicals from reaching the pipe surface.

   a. stabilization, a slime layer
   b. polyphosphates, an acid
   c. polyphosphates, numerous massive bacterial colonies
   d. stabilization, numerous massive bacterial colonies

9. Hypochlorous acid is favored over hypochlorite ion when the pH is __________ and the temperature is __________.

   a. low, high
   b. high, high
   c. low, low
   d. high, low

10. What filter has an unstratified media distribution?

   a. Slow sand filters
   b. Rapid sand filters
   c. High-rate filters
   d. Dual-media filters

11. A well produces 97 gpm. If the drawdown for the well is 22 ft, what is the specific yield in gpm/ft?

   a. 4.4 gpm/ft
   b. 2,134 gpm/ft
   c. 0.23 gpm/ft
   d. 440 gpm/ft

12. What is (are) the combined chlorine residual(s) at pH values ranging from 6.5 to 8.5?

   a. Monochloramine
   b. Dichloramine
   c. Monochloramine and dichloramine
   d. Dichloramine and trichloramine

13. If the total hardness of a water sample is 97 mg/L as $CaCO_3$ and the calcium hardness is 81 mg/L as $CaCO_3$, what is the magnesium hardness?

    a. 14.5 mg/L Mg hardness as $CaCO_3$
    b. 14 mg/L Mg hardness as $CaCO_3$
    c. 15 mg/L Mg hardness as $CaCO_3$
    d. 16 mg/L Mg hardness as $CaCO_3$

14. When iron alone is treated using manganese greensand filters, the chlorine feed rate should be about __________.

    a. 1.0 to 1.2 times the iron concentration
    b. 1.2 to 1.5 times the iron concentration
    c. 1.5 to 1.8 times the iron concentration
    d. 1.5 to 2.0 times the iron concentration

15. Chlorination of water that has high concentrations of natural organic matter will produce

    a. PCBs
    b. BTEX
    c. MTBX
    d. THMs

16. When disinfecting with chloramines, when can the ammonia be added to the process?

    a. Before the chlorine
    b. Same time as the chlorine
    c. After the chlorine
    d. All of the above

17. What will settle out of the water in a short period of time?

    a. Protozoans
    b. Organic matter
    c. Very fine silt
    d. Bacteria

18. Point-of-use activated carbon filters are most effective for removing __________.

    a. volatile organic chemicals
    b. microorganisms
    c. radon
    d. chlorine

19. Potassium permanganate is used as an oxidant for control of __________.

    a. iron and manganese
    b. organic precursors
    c. tastes and odors
    d. All of the above

20. The USEPA Stage 1 Disinfectants and Disinfection Byproduct Rule established maximum contaminant levels for haloacetic acids of __________.

    a. 40 µg/L
    b. 50 µg/L
    c. 60 µg/L
    d. 80 µg/L

21. What must an operator periodically check regarding the cyclone separator system?
    a. Oil for cyclone shift
    b. Storage bin or hopper to empty of sand and grit
    c. Reject water quantity and adjust appropriately
    d. Flow of reject water

22. These types of chemicals chemically tie up scale-forming ions and are eventually ingested by consumers.
    a. Coagulants
    b. Coagulant aids
    c. Sequestering agents
    d. Water-softening chemicals

23. An operator should record filter __________ measurements, condition, and depth of media on a __________ basis.
    a. bed expansion, weekly
    b. sieve analysis, monthly
    c. bed expansion, quarterly
    d. sieve analysis, annual

24. Lime will precipitate noncarbonate hardness from the water, but it also forms noncarbonate salts that cause noncarbonate hardness. What are these salts?
    a. $Na_2SO_4$
    b. $CaSO_4$
    c. $CaCl_2$
    d. Both b and c

25. What are absolute ratings based on with regard to membranes?
    a. Largest pore opening
    b. Average pore opening
    c. Smallest pore opening
    d. The standard deviation from the average pore openings

26. How are manganese greensand beds regenerated?
    a. Backwashed with finished water containing a solution of manganese oxide
    b. A solution of manganese oxide is applied through the inlet.
    c. Backwashed with permanganate solution
    d. Air-backwash, then desired dose of permanganate solution applied at the inlet

27. Find the detention time in minutes for a clarifier that has a diameter of 150.0 ft and a water depth of 14.5 ft if the flow rate is 8.48 mgd.
    a. 325 min
    b. 325.3 min
    c. 326 min
    d. 330 min

28. The most effective process for removing radon gas, when considering cost and footprint, would be __________.

    a. packed tower aeration
    b. granular activated carbon
    c. ion exchange
    d. diffused aeration

29. What is the primary advantage of placing a properly sized anthracite coal layer over manganese greensand?

    a. Greater iron-holding capacity by the greensand
    b. Longer filter run times
    c. Higher degree of coal and greensand mixing, aiding adsorption
    d. Less floc penetration into the greensand; thus, more iron sticks to the greensand

30. Convert 39.7°F to degrees Celsius.

    a. 4.3°C
    b. 103.5°C
    c. 13.86°C
    d. 54.1°C

31. What factor will cause problems for chlorination?

    a. pH of 6.8
    b. Long contact time
    c. Foreign substances
    d. High concentration

32. In the lime–soda ash softening process, smaller plants find that __________ is less expensive to purchase; regardless of the size of the plant, __________ needs to be mixed long enough because it dissolves rather slowly.

    a. calcium oxide, lime
    b. calcium hydroxide, lime
    c. sodium carbonate, soda ash
    d. sodium bicarbonate, soda ash

33. What does the optimal pH establish when applied to iron and manganese oxidation?

    a. Detention time
    b. Percentage oxidation
    c. Charge on the precipitate
    d. Required quantity of chemical oxidant

34. A softener unit has 102 $ft^3$ of resin with a capacity of 24.8 kilograins/$ft^3$. How many gallons of water will the unit treat if the water contains 20.6 gpg?

    a. 0.01 gal
    b. 123 gal
    c. 243 gal
    d. 123,000 gal

35. A water plant using chloramines as a disinfectant has allowed sizable amounts of turbidity to enter the distribution system. If this occurs over a period of years, what will occur at the far end of the distribution system?

    a. Taste and odor will develop
    b. Bacterial growth will be supported
    c. Nitrification
    d. All of the above

36. The reduction of color would most likely occur after __________.

    a. coagulation, flocculation, and sedimentation
    b. softening
    c. filtration
    d. adsorption

37. To eliminate hydrogen sulfide ($H_2S$), the pH of the water should be lowered to ____ or less before beginning aeration.

    a. 12.0
    b. 7.0
    c. 6.0
    d. 8.0

38. Permanganate reactions are largely dependent on __________.

    a. temperature
    b. pH
    c. total organic carbon
    d. alkalinity

39. What are the primary reasons that more small water systems are installing pressure filters?

    a. Because they are easy to operate and maintain
    b. Because installation and operating costs are low
    c. Because of their small footprint and low energy usage
    d. Because water production is fast and footprint is small

40. Because lime slurries have a __________ and a __________ in water, calcium carbonate will __________ on anything the slurry touches.

    a. high pH, low solubility, precipitate
    b. high pH, high alkalinity, absorb
    c. high conductivity, high Langelier saturation index (LSI), adhere
    d. high conductivity, low LSI, absorb

41. Convert 30.7 mgd into cubic feet per second ($ft^3/s$).

    a. 14.1 $ft^3/s$
    b. 47.5 $ft^3/s$
    c. 355.3 $ft^3/s$
    d. 2657.6 $ft^3/s$

42. It is recommended that the free chlorine residual at a customer's tap should be at least __________.

    a. a trace
    b. 0.1 mg/L
    c. 0.2 mg/L
    d. 0.5 mg/L

43. Microfiltration is considered to be intermediate between __________ based on pore size.

    a. ultrafiltration and nanofiltration
    b. nanofiltration and reverse osmosis
    c. ultrafiltration and multimedia granular filtration
    d. multimedia granular filtration and nanofiltration

44. After microscreening, there should be a significant reduction in __________.

    a. turbidity
    b. algae
    c. iron and manganese
    d. undesirable dissolved gases

45. How can the problem of poorly formed floc be solved?

    a. Increase the coagulant dosage
    b. Remove excess baffles
    c. Use a coagulant aid
    d. Clean the sedimentation basin to reduce short-circuiting

46. What chemical used in corrosion control gives off large quantities of $CO_2$ when it comes in contact with an acid?

    a. Sodium silicate
    b. EDTA
    c. Soda ash
    d. Sodium hydroxide

47. Alum requires __________ to be present naturally or added for proper coagulation, and about __________ is required for each milligram of alum added per liter.

    a. carbonate or bicarbonate, 0.5 mg/L
    b. carbonate or bicarbonate, 1.0 mg/L
    c. alkalinity, 0.5 mg/L
    d. alkalinity, 1.0 mg/L

48. A cloudy appearance in the last flocculation chamber is usually caused by __________.

    a. excessive mixing
    b. not enough mixing
    c. inadequate coagulant dosage
    d. too much coagulant aid

49. If a pump discharges 6,720 gal in 1 hr and 52 min, how many gpm is the pump discharging?

    a. 112 gpm
    b. 56.0 gpm
    c. 129 gpm
    d. 60.0 gpm

50. If the water depth in a reservoir is 14.06 ft, what is the pressure at 11.5 ft below the surface in $lb/ft^2$?

    a. 5 $lb/ft^2$
    b. 160 $lb/ft^2$
    c. 718 $lb/ft^2$
    d. 877 $lb/ft^2$

51. What type of meter is most commonly used for customer metering?
    a. Positive-displacement meters
    b. Venturi meters
    c. Velocity meters
    d. Ultrasonic flowmeters

52. If 250 mL of a 70% solution is mixed with 100 mL of water, what is the final concentration?
    a. 20% final concentration
    b. 98% final concentration
    c. 28% final concentration
    d. 50% final concentration

53. In general, the size of floc that has been found to be optimal for settling efficiency is __________.
    a. pinhead-size floc
    b. dime-size floc
    c. nickel-size floc
    d. quarter-size floc

54. In a pulsator clarifier, what is the purpose of the pulses, and how often are pulses generated?
    a. To maintain a uniform sludge layer; 20–25 s
    b. To maintain a uniform sludge layer; 40–50 s
    c. To reduce short circuiting; 30–40 s
    d. To reduce short circuiting, 60–90 s

55. A water plant has four rectangular clarifiers with weir lengths of 85 ft each. What is the total weir overflow rate in gpd/ft if the flow is 2.71 mgd?
    a. 800 gpd/ft
    b. 3,190 gpd/ft
    c. 8,000 gpd/ft
    d. 31,900 gpd/ft

56. UV application will __________.
    a. inhibit bacterial growth and replication
    b. convert nitrates to nitrites
    c. convert bromide to bromines
    d. convert bromide to bromate

57. A filter produced a total of 2,845,200 gallons between backwashes. If the filter is 28.0 ft by 20.0 ft, what is the UFRV gal/ft$^2$?
    a. $2 \times 10^{-4}$ gal/ft$^2$
    b. 5,080 gal/ft$^2$
    c. 101,600 gal/ft$^2$
    d. 142,300 gal/ft$^2$

58. Calcium hypochlorite used in water treatment disinfection ranges in percentage strength of available chlorine from __________.
    a. 60% to 65%
    b. 65% to 70%
    c. 60% to 70%
    d. 55% to 70%

59. Manganese greensand filters remove iron and manganese by a combination of both ________.

   a. oxidation and straining
   b. oxidation and absorption
   c. adsorption and oxidation
   d. adsorption and absorption

60. After water has been softened, it is close to pH 11; thus, it needs to be stabilized by a process known as recarbonation. What small amount of hardness chemical is formed and what is precipitated in this process?

   a. Calcium carbonate, magnesium hydroxide
   b. Calcium bicarbonate, magnesium carbonate
   c. Calcium hydroxide, calcium carbonate
   d. Magnesium hydroxide, calcium carbonate

61. UV is used as a disinfectant because it is effective in the inactivation of ________ and because it __________.

   a. *Giardia,* is inexpensive to use
   b. *Cryptosporidium,* does not produce DBPs
   c. viruses, is inexpensive to use
   d. *Giardia,* has a very small "footprint" at a water plant

62. A common problem with hypochlorinators is __________.

   a. organic precipitates
   b. stuck inlet valve
   c. broken diaphragm
   d. broken ejector

63. If algae or slime growth in sedimentation basins are a problem, drain the basin and apply a solution of __________ per liter of water using a brush to remove the growth.

   a. 5 g of copper sulfate and 5 g of lime
   b. 10 g of copper sulfate and 10 g of soda ash
   c. 10 g of copper sulfate and 10 g of lime
   d. 5 g of copper sulfate and 10 g of soda ash

64. What two major problems will improperly stabilized softened water cause?

   a. Scale in filter and/or pipes
   b. Scale deposits or corrosive water
   c. Scale deposits and/or tastes in finished water
   d. Flow problems and customer complaints

65. Water is flowing through a rectangular channel that is 4.05 ft wide and 1.90 ft deep. If the flow is 8.5 $ft^3/s$, what is the velocity in feet per second (ft/s) of the water?

   a. 0.91 ft/s
   b. 0.99 ft/s
   c. 1.1 ft/s
   d. 7.69 ft/s

66. Coagulation and flocculation are __________ processes.

    a. physical
    b. chemical
    c. electrochemical
    d. chemical and mechanical

67. The $C \times T$ value is calculated using most or part of the treatment plant, some distribution piping, and to the __________.

    a. influent of the first storage tank
    b. effluent of the first storage tank
    c. first customer
    d. furthest point in the distribution system

68. The most important reason to control corrosion and scaling in the distribution system is to __________.

    a. meet regulations
    b. protect public health
    c. improve water quality
    d. extend plumbing equipment life

69. A bacterial slime layer prevents the deposition of __________, and when the slime occasionally gets sloughed away, the slime carries away __________ and leaves the pipe surface bare.

    a. zinc orthophosphate, ferrous hydroxide
    b. zinc orthophosphate, ferric hydroxide
    c. calcium carbonate, calcium carbonate
    d. calcium carbonate, calcium oxide

70. What is the major advantage of GAC over anthracite?

    a. Better adsorption of organics
    b. Slower exhaustion by chlorine compared to anthracite
    c. Heavier than anthracite so there is less loss of media
    d. Longer life than anthracite

71. The success of the conventional treatment design is primarily due to the __________ process.

    a. coagulation
    b. sedimentation
    c. filtration
    d. All of the above

72. How is hardness expressed in the ion exchange process?

    a. grains/mg/L
    b. grains/17.12 mg/L
    c. grains/gal
    d. grains/L

73. A water treatment plant has eight filters with an average flow rate of 5.87 gpm/$ft^2$. If the plant flow is 58 $ft^3$/s, what is the filtration area of each filter?

    a. 550 $ft^2$ for each filter
    b. 553.7 $ft^2$ for each filter
    c. 554 $ft^2$ for each filter
    d. 560 $ft^2$ for each filter

74. Manganese will produce _______ water.

    a. red
    b. brown or black
    c. brown
    d. black and gray

75. Air binding is caused by pressure in the filters __________ than atmospheric pressure, and it can cause __________ flow.

    a. higher, resistance to
    b. lower, resistance to
    c. higher, increase in
    d. lower, increase in

76. Determine the hydrated lime dose required in mg/L for water with the following characteristics:

| | Source Water | Softened Water before Blending |
|---|---|---|
| Total alkalinity, mg/L | 187 mg/L as $CaCO_3$ | 32 mg/L |
| Total hardness, mg/L | 361 mg/L as $CaCO_3$ | 62 mg/L |
| $CO_2$, mg/L | 9.85 mg/L | 0 mg/L |
| $Mg^{2+}$ | 29 mg/L | 8.7 mg/L |
| pH | 7.6 | 8.5 |
| Lime purity | 90.0% | |

Use the following table to answer this question:

Calculations for chemical precipitation softening process

| Molecular weights of chemical compounds | |
|---|---|
| Compound | Molecular Weight |
| Alkalinity, as $CaCO_3$ | 100.1 |
| Carbon dioxide, $CO_2$ | 44.0 |
| Hardness, as $CaCO_3$ | 100.1 |
| Hydrated lime, $Ca(OH)_2$ | 74.1 |
| Magnesium, $Mg^{2+}$ | 24.3 |
| Magnesium hydroxide, $Mg(OH)_2$ | 58.3 |
| Quicklime, CaO | 56.1 |
| Soda ash, $Na_2CO_3$ | 106.0 |

a. 215 mg/L of $Ca(OH)_2$
b. 220 mg/L of $Ca(OH)_2$
c. 247 mg/L of $Ca(OH)_2$
d. 247.3 mg/L of $Ca(OH)_2$

77. What filter types would be best for direct filtration?

a. Slow sand filters
b. Monomedium filters
c. Rapid sand filters
d. High-rate filters

78. What are the two broad classes of corrosion in water systems?

a. Localized corrosion and uniform corrosion
b. Galvanic corrosion and concentration cell corrosion
c. Precipitation corrosion and electrochemical corrosion
d. Concentration cell corrosion and uniform corrosion

79. The oxidation process using chlorine will also consume __________.

a. ammonia
b. organics
c. alkalinity
d. iron and manganese

80. Sodium hypochlorite is usually purchased at 12% to 12.5% and is then diluted down to a more stable percentage of __________.

a. 8%
b. 6%
c. 5%
d. 4%

81. Disinfection chemicals must conform to standards set by __________.

a. USEPA and ANSI
b. ANSI and AWWA
c. NSF International
d. USEPA and AWWA

82. If a particular section of the distribution system suddenly has a drop in chlorine residual, what most likely has occurred?

a. A cross-connection
b. Loss of chlorine feed at the plant
c. Broken main in the area
d. Closed valve in the area

83. What type of automation control would work best if the chlorine demand varies but the flow is constant?

a. Residual control
b. Combined flow and residual control
c. Start–stop control
d. Proportional pacing

84. Point-of-use activated carbon filters are not usually effective in removing __________.

   a. radon
   b. synthetic organic chemicals
   c. inorganic chemicals
   d. tastes and odors

85. Where is the heater located that vaporizes any chlorine liquid that may develop downline from an online one-ton container?

   a. After the one-ton container's regulator
   b. Before the chlorinator
   c. After the ejector
   d. Before the vacuum shutoff valve

86. In the oxidation of iron and manganese, the pH must be adjusted to an optimal level. What must be determined to accomplish the optimal pH?

   a. The zeta potential
   b. Solubility of the precipitate
   c. Charge of the precipitate
   d. All of the above

87. Iodine as a disinfectant would most likely be used at _________.

   a. campgrounds
   b. schools
   c. large public office buildings
   d. Northwest United States area universities

88. Mechanical mixers should never be used in conjunction with __________.

   a. resins
   b. strong acids or bases
   c. in-line coagulation
   d. saturators

89. After iron and manganese precipitates are formed by the oxidation process, they are usually removed by _________.

   a. sequestration
   b. sedimentation
   c. flocculation
   d. filtration

90. A conventional water treatment plant has six flocculators with a detention time of 1 hr and 15 min. What should the operator do for the best results?

   a. Increase the flow
   b. Increase the mixing speed of the flocculators to prevent premature settling of the floc
   c. Turn off the last flocculator
   d. Turn off the last three flocculators

91. To precipitate calcium carbonate, a pH of about __________ is necessary.

    a. 8.5
    b. 9.1
    c. 9.4
    d. 10.6

92. If 4.58 oz of polymer are mixed into exactly 10.0 gal of water, what is the percentage of polymer in the water?

    a. 0.342% of polymer
    b. 0.343% of polymer
    c. 0.384% of polymer
    d. 0.40% of polymer

93. A conventional treatment plant processes 4,475 gpm. If the lime dosage is 121 g/min, how many lb of lime will the plant use in one month (30 days)? What is the dosage in mg/L?

    a. 11,500 lb/month of lime; 7.14 mg/L of lime
    b. 11,510 lb/month of lime; 7.20 mg/L of lime
    c. 11,520 lb/month of lime; 7.2 mg/L of lime
    d. 11,520 lb/month of lime; 7.19 mg/L of lime

94. What is the disadvantage of direct filtration compared to conventional treatment?

    a. It must be carefully monitored to prevent turbidity breakthrough
    b. Longer process times
    c. More coagulant aid
    d. Much more coagulant chemicals are needed.

95. Potassium permanganate is a __________.

    a. weak oxidizer
    b. strong oxidizer
    c. weak base
    d. strong base

96. If a given compound is formed by the combination of elements, the proportion, by weight, of the element/s making up the compound is always the same regardless of the method used to form the compound defines the __________.

    a. Law of Definite Proportions
    b. Law of Multiple Proportions
    c. Law of Combining Weight
    d. Both a and b

97. The term "schmutzdecke" is affiliated with which type of filter system?

    a. Membrane filtration
    b. Slow sand filtration
    c. Rapid sand filters
    d. High-rate filters

98. Weak floc is usually due to __________.

    a. inadequate mixing in the rapid-mix or flocculation basins
    b. water temperature being too cold
    c. inadequate recycling of settled sludge
    d. incorrect streaming current monitor set point

99. What is the surface loading rate for a sedimentation basin that is 175 ft by 52.3 ft if it is treating an instantaneous flow rate of 17.1 $ft^3/s$?

a. 9,152 $gpd/ft^2$
b. 1,183 $gpd/ft^2$
c. 1,207 $gpd/ft^2$
d. 1,210 $gpd/ft^2$

100. What is the backwash rate in $gpm/ft^2$ given the following:

Filter is 32 ft in length by 24 ft in width.

Backwash flow is 30.5 $ft^3/s$.

a. 17.8 $gpm/ft^2$
b. 17.9 $gpm/ft^2$
c. 18 $gpm/ft^2$
d. 19 $gpm/ft^2$

101. What is an operating factor that would affect floc formation?

a. Weak floc
b. Slow floc formation
c. Low water temperature
d. Inadequate flocculation time

102. Which of the following processes is NOT combined in a solids-contact basin?

a. Coagulation
b. Flocculation
c. Reverse osmosis
d. Sedimentation

103. What is the patented filter underdrain system called that uses a series of perforated vitrified clay blocks with channels to carry the water?

a. Wheeler bottom
b. Leopold filter bottom
c. Porous channel filter bottom
d. Schultz filter bottom

104. What level does the pH need to be raised to precipitate magnesium hydroxide?

a. 8.3
b. 9.8
c. 10.6
d. 11.3

105. A fluoride dose of 1.20 mg/L is desired for treating a flow of 1,285 gpm. How many lb/day of sodium silicofluoride ($Na_2SiF_6$) with a commercial purity of 98% and a fluoride ion content of 60.5% is required? The water being treated contains 0.08 mg/L fluoride.

a. 29 lb/d of $Na_2SiF_6$
b. 29.2 lb/d of $Na_2SiF_6$
c. 31 lb/d of $Na_2SiF_6$
d. 31.2 lb/d of $Na_2SiF_6$

106. How is magnesium noncarbonate hardness removed?

   a. With lime
   b. With soda ash
   c. With lime then soda ash
   d. With soda ash then lime

107. Comparisons of filters should not be based on __________, because they vary from manufacturer to manufacturer, but should be based on __________.

   a. nominal ratings, pore ratings
   b. nominal ratings, absolute ratings
   c. pore ratings, nominal ratings
   d. absolute ratings, nominal ratings

108. The excess lime-softening process is used so that __________ will precipitate out of the water.

   a. calcium nitrate
   b. calcium chloride
   c. magnesium hydroxide
   d. magnesium sulfate

109. __________ is used occasionally instead of sodium carbonate because it will __________ without as much an increase in __________.

   a. Lime, increase alkalinity, pH
   b. Soda ash, increase pH, alkalinity
   c. Sodium bicarbonate, increase alkalinity, pH
   d. Caustic, increase pH, alkalinity

110. What type of pretreatment must be cleaned manually when the amount of debris is relatively small?

   a. Bar screen
   b. Sand trap
   c. Wire-mesh screen
   d. Cyclone degritter

111. After screening and presedimentation, there should be a significant reduction in __________.

   a. sediment and other debris
   b. turbidity
   c. algae
   d. iron and manganese

112. Carbon dioxide is a __________, __________ gas that is __________.

   a. colorless, odorless, lighter than air
   b. poisonous, heavier than air, very dangerous
   c. colorless, odorless, heavier than air
   d. colorless, poisonous, able to cause asphyxiation under the right conditions

113. Why would there still be taste and odor present in water that has been treated with PAC?

a. Aeration too short, PAC overwhelmed
b. Coagulation and flocculation insufficient
c. Preoxidation insufficient
d. PAC and chlorine added too close together

114. What corrosion prevention chemical may be the best overall treatment approach?

a. Lime
b. Phosphate
c. Silicate
d. Sodium hydroxide

115. Preoxidation has been shown to improve _________.

a. aeration and flocculation
b. sedimentation
c. flocculation and sedimentation
d. flocculation, sedimentation, and filtration

116. Find the specific yield in gpm/ft if a well produces 49 gpm and the drawdown for the well is 11.7 ft.

a. 0.24 gpm/ft
b. 2.4 gpm/ft
c. 4.2 gpm/ft
d. 0.41 gpm/ft

117. If excessive algae grow in a presedimentation impoundment, what is the best solution?

a. Line the bottom to limit nutrients for the algae to grow
b. Cover the impoundment
c. Chlorinate to kill the algae
d. Use potassium permanganate to reduce the number of algae

118. What is another name for caustic soda, and what is its chemical formula?

a. Sodium hydroxide, $NaOH$
b. Sodium bicarbonate, $Na_2CO_3$
c. Sodium carbonate, $Na_2CO_3$
d. Sodium bicarbonate, $NaHCO_3$

119. What chemical can cause severe scaling if not properly stabilized?

a. Lime
b. Caustic
c. Soda ash
d. Sodium bicarbonate

120. How many pounds of 65.0% calcium hypochlorite is required for a 25 mg/L dosage for a tank that is 100.1 ft in diameter and has a water level of 3.25 ft?

a. 25.9 lb of calcium hypochlorite
b. 26 lb of calcium hypochlorite
c. 61 lb of calcium hypochlorite
d. 61.3 lb of calcium hypochlorite

121. Which of the following chemicals looks like broken glass?

   a. Sodium zinc phosphate
   b. Sodium hexametaphosphate
   c. Zinc orthophosphate crystals
   d. Orthophosphate crystals

122. What is conventional treatment?

   a. Flash mixing, coagulation, flocculation, and filtration
   b. Coagulation, flocculation, sedimentation, and filtration
   c. Coagulation, flocculation, sedimentation, filtration, and chlorination
   d. Flash mixing, coagulation, flocculation, sedimentation, filtration, and chlorination

123. What type of flash-mixing process gives the best mixing and control?

   a. Static mixers
   b. Mechanical mixers
   c. Baffled chambers
   d. Pumps

124. Chloramines are weak at inactivating certain __________.

   a. viruses
   b. coliforms
   c. fungi
   d. blue-green bacteria

125. Manganese greensand filtration is used to remove __________.

   a. manganese, arsenic, and sulfates
   b. manganese, iron, and arsenic
   c. manganese, iron, and hydrogen sulfide
   d. manganese, hydrogen sulfide, and other sulfates

126. The application of chlorine before a microstrainer creates many problems, but it will not ________.

   a. accentuate tastes and odors
   b. make it difficult to clean off dead algae
   c. corrode the stainless-steel mesh
   d. cause the oxidation of iron, which will form a jelly-like coating

127. What noncarbonate hardness chemical is produced when sulfuric acid is used to stabilize water?

   a. $CaSO_4$
   b. $MgSO_4$
   c. $Na_2SO_4$
   d. $K_2SO_4$

128. The USEPA Stage 1 Disinfectants and Disinfection Byproduct Rule established Maximum Residual Disinfectant Levels for chloramine of _________.

   a. 3 mg/L
   b. 4 mg/L
   c. 5 mg/L
   d. 6 mg/L

129. What is a patented filter underdrain system called that uses small porcelain spheres of various sizes in conical depressions?

 a. Wheeler bottom
 b. Leopold filter bottom
 c. Schultz filter bottom
 d. Ball filter bottom

130. Find the motor horsepower for a pump station with the following parameters:

 Motor efficiency: 88%

 Pump efficiency: 74%

 Total head (TH): 109 ft

 Flow: 2.42 mgd

 a. 30 mhp
 b. 53 mhp
 c. 63 mhp
 d. 71 mhp

131. Once a slow sand filter has been cleaned, it may be necessary to filter for waste for as long as _________.

 a. 4 hours
 b. 8 hours
 c. 1 day
 d. 2 days

132. The lime-softening process used in removing iron works in the pH range of _________.

 a. 7–9
 b. 8–9
 c. 9–10
 d. 10–11

133. Greensand grains are typically _________ than silica sands, so the _________ can quickly become _________.

 a. smaller, head loss, excessive under a heavy load
 b. denser, compacted grains, excessively clogged with precipitates and other filterable material
 c. smaller, compacted grains, excessively clogged with precipitates and other filterable material
 d. denser, head loss, excessive under high flow rates

134. What pressure differential breaks up both the $MnO_2(s)$ and the glauconite mineral sand grain, which causes fines that reduce filter run times?

 a. 8 psi
 b. 10 psi
 c. 12 psi
 d. 14 psi

135. What type of underdrain system may or may not have a gravel pack at its base?

a. Porous plate bottom
b. Wheeler filter bottom
c. False floor system
d. Perforated tile bottom

136. Water is flowing at a velocity of 1.68 ft/s in a 12.0-in.-diameter pipe. If the pipe changes from the 12.0-in. to an 8.0-in. pipe, what will be the velocity in the 8-in. pipe?

a. 1.12 ft/s in the 8.0-in. pipe
b. 2.52 ft/s in the 8.0-in. pipe
c. 3.8 ft/s in the 8.0-in. pipe
d. 3.87 ft/s in the 8.0-in. pipe

137. An effective way to remove __________ is through __________.

a. *Cryptosporidium* oocysts, coagulation and filtration
b. *Cryptosporidium* oocysts, filtration and chlorine disinfection
c. *Giardia* cysts, coagulation
d. *Giardia* cysts, sedimentation

138. Membranes made of Teflon® (polytetrafluoroethylene) have __________ oxidant resistance.

a. poor
b. moderate
c. good
d. high

139. When the $CO_2$ level goes above 5 mg/L to 15 mg/L as $CO_2$, it makes it more difficult to remove __________.

a. radon
b. tastes and odors
c. organic acids
d. iron

140. The most preferable method of removing manganese is by using _________.

a. chlorine
b. aeration
c. softening
d. potassium permanganate

141. If tuberculation is already present, then adding __________ can actually increase the problems because these chemicals can loosen the deposits and cause them to break away.

a. soda ash
b. sodium carbonate
c. sodium bicarbonate
d. polyphosphates

142. Hypochlorous acid is a __________ acid that __________ and kills bacteria.

a. weak, inactivates *Cryptosporidium*
b. strong, easily penetrates
c. weak, easily penetrates
d. strong, inactivates *Cryptosporidium*

143. The most important factor(s) in chlorination is (are) __________.

a. pH of the water and concentration of the chlorine
b. contact time between chlorine and the water and foreign substances in the water
c. concentration of the chlorine and contact time between chlorine and the water
d. pH of the water and foreign substances in the water

144. A 10-min drawdown test result showed that 14 mL of polymer aid was being used to help treat raw water. The specific gravity of the polymer aid is 1.25. If the plant is treating 5,450 gpm, what is the polymer dosage in mg/L?

a. 0.068 mg/L of polymer aid
b. 0.085 mg/L of polymer aid
c. 0.095 mg/L of polymer aid
d. 0.096 mg/L of polymer aid

145. Calculate the volume in cubic feet for a pipeline that is 16.0 in. in diameter and 862 ft long.

a. 1,805 $ft^3$
b. 1,200 $ft^3$
c. 3,610 $ft^3$
d. 21,650 $ft^3$

146. Iron bacteria in a distribution system pipe can increase corrosion if enough dissolved iron is present. What do the iron bacteria produce that increases the corrosion rate?

a. $CO_2$
b. $H_2S$
c. Organic acids
d. Sulfate ions ($SO_4^{-2}$)

147. Why should an operator NOT add extra, untested material to a multimedia filter?

a. Specific gravity may be different.
b. Roundness or angularity of the media may be different.
c. Because the effective size may be different
d. Only A and C

148. What must the effluent ntu for a slow sand filter be kept at during operation?

a. Less than 0.3 ntu
b. Equal to 0.3 ntu or less
c. Less than 1.0 ntu
d. Equal to 5.0 ntu or less

149. What is an advantage for installing a pressure filter for a small water system?

a. Very low maintenance
b. Low energy consumption
c. Air binding will not occur
d. Media is not lost

150. What type of instrument will help operators improve the coagulation process the most?

a. Turbidity meters at raw water, sedimentation basin, and filters
b. Accurate Venturi or magnetic flow meters
c. Streaming current monitors
d. Particle counters

151. The pH range for a manganese greensand filter for removing iron and manganese should be adjusted to a range of _________.

 a. 7.5–8.0
 b. 6.8–7.5
 c. 6.2–6.8
 d. 6.2–6.5

152. The best and most reliable way to detect small chlorine leaks is by __________. These small leaks are only __________ or less chlorine in air.

 a. smell, 3 ppm
 b. ammonia, 1 ppm
 c. a chlorine leak detector, 1 ppm
 d. finding discoloration, 5 ppm

153. How long will it take in hours and minutes to unload a tanker truck filled with 14,100 L of liquid alum if the truck's pump unloads the alum at 80 gpm? The tank's capacity is 10,000 gal, and it has about 3,500 gal of alum in it.

 a. 1 hour 30 min
 b. 47 min
 c. 2 hours 56 min
 d. 11 hours 7 min

154. Iron and manganese in the distribution system can lead to bacterial slimes. What is a disadvantage of the elements being in the distribution system?

 a. Reduced pipeline flow
 b. Clogged meters and valves
 c. Corrosion of metal pipes
 d. Increased chlorine demand

155. Chloramine is a __________ and is not especially effective against certain __________.

 a. moderate oxidant, viruses
 b. moderate oxidant, bacterial pathogens
 c. weak oxidant, viruses
 d. weak oxidant, bacterial pathogens

156. In general, what is the AWWA recommended minimum free chlorine residual at the extremities of the distribution system?

 a. 0.05 mg/L
 b. 0.1 mg/L
 c. 0.2 mg/L
 d. 0.25 mg/L

157. There are a number of techniques and technologies to introduce ozone into water. The most commonly used contactor is the __________.

 a. fine-bubble diffuser
 b. spray chamber
 c. turbine mixer
 d. packed columns

158. __________ is defined as the color of water from which __________ has been removed, and __________ is color caused by __________.

a. True color, suspended matter, apparent color, turbidity
b. True color, turbidity, apparent color, suspended matter
c. Apparent color, suspended matter, true color, turbidity
d. Apparent color, turbidity, true color, suspended matter

159. In the flocculation basins, as the floc grows in size they become ____________, so the mixing must be ___________ so that the floc ____________.

a. stronger, greater, grows larger
b. stronger, greater, does not settle
c. fragile, slowed, does not shear
d. weaker, stopped, grows larger

160. Monomedium filter media is __________ than a conventional filter because the media is __________.

a. shallower, finer
b. deeper, coarser
c. shallower, coarser
d. deeper, finer

161. Surface water can only be made consistently safe if it is disinfected after removing __________.

a. bacterial and viral pathogens
b. *Giardia* and *Cryptosporidium*
c. the turbidity
d. viruses and raising the pH to 7.8 to kill all other pathogens

162. Tastes and odors caused by oils produced by certain algae may be effectively removed by _________.

a. chemical oxidants
b. diffused-bubble aerator
c. packed tower aeration
d. particle filtration

163. Filtration alone can be used after oxidation of iron and manganese if the concentration of the metals is below about _________.

a. 5 mg/L
b. 8 mg/L
c. 10 mg/L
d. 15 mg/L

164. It is recommended that an operator record the results of solids retention analysis before and after backwashing on a __________ basis.

a. monthly
b. quarterly
c. biannual
d. yearly

165. What chemical would work best for coating distribution pipes if the calcium concentration were 2 mg/L as $CaCO_3$?

a. Caustic
b. Polyphosphates
c. Lime
d. Soda ash

166. Examples of advantages and disadvantages for disinfectants include __________, which adds an objectionable taste to the water for some people; __________ has a persistent residual; and __________ will only slightly be affected by pH.

a. chloramines, chlorine dioxide, chlorine
b. chlorine dioxide, chloramines, ozone
c. chloramines, chlorine, chlorine dioxide
d. ozone, chlorine dioxide, chloramines

167. The most common corrosion inhibitors are __________.

a. polyphosphates and silicates
b. sodium carbonate and sodium bicarbonate
c. lime with sodium carbonate or sodium bicarbonate
d. polyphosphates and lime

168. UV reaction with microorganisms is purely __________ and therefore not significantly affected by __________.

a. physical, pH
b. chemical, temperature
c. physical, high alkalinity
d. chemical, high organics

169. When water is added too fast to this treatment chemical, it will generate heat and splattering of the chemical.

a. Sodium silicate
b. Sulfuric acid
c. Potassium permanganate
d. Sodium bicarbonate

170. After breakpoint chlorination, what may remain besides the free chlorine?

a. Ammonia
b. Chloroorganics
c. Iron and manganese
d. Ammonia and chloramines

171. Calculate the feed rate for hydrofluosilicic acid in gpd given the following data:

| | |
|---|---|
| Flow rate is 15.5 mgd | Treated with 21% solution of $H_2SiF_6$ |
| Fluoride desired is 1.0 mg/L | Fluoride ion percentage is 79% |
| Fluoride in raw water is 0.12 mg/L | $H_2SiF_6$ weighs 9.83 lb/gal |

a. 69.8 gpd of $H_2SiF_6$
b. 70 gpd of $H_2SiF_6$
c. 88.8 gpd of $H_2SiF_6$
d. 89 gpd of $H_2SiF_6$

172. Dry chemicals should be in solution before mixing them with the water to be treated. What water is recommended to mix the dry chemical in some cases to avoid precipitates?

a. Deionized water
b. Distilled water
c. Softened water
d. All of the above

173. When a backwash is occurring, what should the operator do?

a. Hose down the walls
b. Sample the floc
c. View rate of rise
d. All of the above

174. How many pounds of 12.5% sodium hypochlorite (NaOCl) are there in 1 gal if the solution weighs 10.0 lb/gal?

a. 1.25 lb/gal of NaOCl
b. 1.3 lb/gal of NaOCl
c. 1.5 lb/gal of NaOCl
d. 1.75 lb/gal of NaOCl

175. When calculating the $C \times Ts$ for a surface water treatment plant, besides needing to know the concentration of the disinfectant and the contact time, what else is needed for the calculation?

a. Average flow rate, temperature, and pH
b. Highest flow rate, pH, and temperature
c. Average hourly flow rate, lowest contactor volume, pH, and temperature
d. Highest hourly flow rate, lowest contactor volume, pH, and temperature

176. Presedimentation basins can be converted with a retrofit of ____________ and this can help ______________.

a. tube settlers, eliminate water fowl problem
b. baffles, reduce aquatic plants
c. lamella plates, control the growth of algae
d. lamella plates, reduce chemical usage

177. What plant-operating factor could make a difference in the proper development of floc?

a. Inadequate flocculation time
b. High turbidity in settled water
c. Aeration reducing calcium carbonate too much
d. Prechlorine dose too high

178. Water is flowing at a velocity of 3.28 ft/s in a 10.0-in.-diameter pipe. If the pipe changes from the 10.0-in. to a 14.0-in. pipe, what will be the velocity in the 14.0-in. pipe?

a. 1.23 ft/s in the 14.0-in. pipe
b. 1.67 ft/s in the 14.0-in. pipe
c. 4.59 ft/s in the 14.0-in. pipe
d. 6.43 ft/s in the 14.0-in. pipe

179. What is glauconite used for in the water industry?

a. Material used to make diffusers mostly for filters
b. Trace compound added to iron pipe along with carbon to make it stronger
c. Naturally occurring material that makes up greensand
d. A type of coal used to make powdered activated carbon

180. The density of an unknown liquid is 70.4 lb/ft$^3$. What is the specific gravity (sp gr) of the liquid?

a. 0.07 sp gr
b. 0.89 sp gr
c. 1.13 sp gr
d. 1.16 sp gr

181. Determine the hydrated lime dose required in mg/L for water with the following characteristics:

| | Source Water | Softened Water |
|---|---|---|
| Total alkalinity, mg/L | 192 mg/L as $CaCO_3$ | 38.7 mg/L as $CaCO_3$ |
| Total hardness, mg/L | 237 mg/L as $CaCO_3$ | 48 mg/L as $CaCO_3$ |
| $CO_2$, mg/L | 6.8 mg/L | 0 mg/L |
| $Mg^{2+}$, mg/L | 19.6 mg/L | 8.8 mg/L |
| pH | 7.3 | 7.9 |
| Lime purity | 92% | |

Use the following table to solve this problem:

Calculations for chemical precipitation softening process

| Molecular weights of chemical compounds | |
|---|---|
| Compound | Molecular Weight |
| Alkalinity, as $CaCO_3$ | 100.1 |
| Carbon dioxide, $CO_2$ | 44.0 |
| Hardness, as $CaCO_3$ | 100.1 |
| Hydrated lime, $Ca(OH)_2$ | 74.1 |
| Magnesium, $Mg^{2+}$ | 24.3 |
| Magnesium hydroxide, $Mg(OH)_2$ | 58.3 |
| Quicklime, CaO | 56.1 |
| Soda ash, $Na_2CO_3$ | 106.0 |

a. 197 mg/L of $Ca(OH)_2$
b. 197.3 mg/L of $Ca(OH)_2$
c. 198 mg/L of $Ca(OH)_2$
d. 200 mg/L of $Ca(OH)_2$

182. The USEPA Stage 1 Disinfectants and Disinfection Byproduct Rule established maximum contaminant levels for total TTHMs of _________.

a. 40 μg/L
b. 50 μg/L
c. 60 μg/L
d. 80 μg/L

183. A filter bed is usually completely fluidized at just __________ expansion.

a. 7%
b. 12%
c. 15%
d. 20%

184. Filters will remove particulate matter by _________.

a. chemical and mechanical straining and absorption
b. absorption and adsorption
c. mechanical straining and adsorption
d. chemical straining and absorption

## Laboratory Analysis

1. What test is done by a laboratory to determine data quality?

a. Performance evaluation
b. Spiked samples to test
c. Quality control test
d. Both b and c

2. A beaker is primarily used for _________.

a. mixing chemicals
b. titrations
c. accurately measuring liquid chemicals
d. reaction vessel and jar testing

3. The Curie unit is the activity of 1 g of _________ or _________ disintegrations per second.

a. uranium, $3.2 \times 10^{10}$
b. uranium, $3.7 \times 10^{9}$
c. radium, $3.7 \times 10^{10}$
d. radium, $3.2 \times 10^{9}$

4. What is the primary constituent of the pH-causing ions in a water sample if the pH is above 10?

a. $HCO_3^{-1}$
b. $OH^{-1}$
c. $CO_2$
d. $CO_3^{-2}$

5. Coliform bacteria will produce gas in __________ broth at __________ within __________.
   a. lactose or brilliant bile tryptose, 35°C, 24–48 hours
   b. lactose or brilliant bile tryptose, 35°C, 48–72 hours
   c. lactose or lauryl tryptose, 35.5°C, 24–48 hours
   d. lactose or lauryl tryptose, 35.5°C, 18–24 hours

6. The number of samples under the Total Coliform Rule that a water system must collect depends on the ________.
   a. number of connections
   b. number of customers it serves
   c. amount of water produced
   d. size of distribution

7. Phenolphthaelin acidity is the same as _________.
   a. $CO_2$
   b. $OH^-$
   c. $HCO_3$
   d. $CO_3$

8. Drinking water samples to be tested for total coliforms should be tested the same day they are collected, but they can be refrigerated for __________ before analysis.
   a. 4 hours
   b. 8 hours
   c. 10 hours
   d. 12 hours

9. What type of laboratory equipment is highly resistant to heat, sudden temperature change, and chemical attack?
   a. Pyrex equipment
   b. Kimax
   c. Porcelain dishes
   d. Beakers

10. Two metals that are typically toxic are __________, and two metals that are not normally toxic are __________.
    a. chromium and cadmium, manganese and iron
    b. cadmium and manganese, chromium and iron
    c. cadmium and iron, chromium and manganese
    d. chromium and manganese, cadmium and iron

11. The total hardness and the total alkalinity of water are used to calculate the _________.
    a. Langelier saturation index
    b. hydroxide, carbonate, and bicarbonate hardness
    c. lime–soda ash dosage
    d. carbonate and noncarbonate hardness

12. What group of organisms are measured that, if present, might indicate the presence of pathogens?

    a. *Cryptosporidium* oocysts
    b. *Coliform* bacteria
    c. *Bacilli* bacteria
    d. *Lactobacillus* bacteria

13. Alkalinity is defined as _________.

    a. the quantity of $CaCO_3$ and noncarbonate hardness that is in the water
    b. the quantity of $CaCO_3$ and $MgCO_3$ that is in the water
    c. a measurement of the water's capacity to neutralize an acid
    d. a measurement of the water's capacity to neutralize an acid to a pH of 4.8

14. What should be done with the lid while collecting a bacteriological sample?

    a. Place the lid with the threads facing down on a clean surface, such as a paper towel
    b. Set the lid up on a clean surface
    c. Hold the cap, threads down, while collecting the sample with the other hand
    d. Put the cap in your pocket so it is not dropped and contaminated

15. The zeta potential is defined as a force that will keep _________ apart and keep them suspended.

    a. salts
    b. ionic molecules
    c. very small clay particles
    d. very small colloidal particles

16. SUVA is an indicator of what disinfection by product precursor substances?

    a. Inorganic and organic content
    b. Humic content
    c. Fulvic content
    d. Protein content

17. Sampling of organic compounds must be done with ultraclean glass vials that have lids made of ________.

    a. polypropylene
    b. polytetrafluoroethylene
    c. high-density polyethylene
    d. polychlorinated biphenyl

18. The best place to collect samples at water treatment plants would most probably be _________.

    a. immediately downstream of chemical addition
    b. worst-case points, such as near settled debris or algae mats
    c. from the main stream flow
    d. near calm water to avoid excessive turbulence

19. What are the chemical formula precipitates when chlorine is added to water containing iron and manganese?

    a. $Fe(OH)_2$ and $MnO$
    b. $Fe(OH)_3$ and $MnO_2$
    c. $Fe(OH)_2$ and $MnO_2$
    d. $Fe(OH)_3$ and $MnO$

20. What is the calcium hardness, as $CaCO_3$, if the water sample has a calcium content of 51 mg/L?

    Use the following equivalent weights table to answer this question.

| Equivalent Weights | |
|---|---|
| Constituent | Equivalent Weight (Eq. Wt.) |
| Calcium, Ca | 20.04 |
| Calcium Carbonate, $CaCO_3$ | 50.045 |
| Magnesium, Mg | 12.15 |

    a. 50.05 mg/L Ca as $CaCO_3$
    b. 70.08 mg/L Ca as $CaCO_3$
    c. 127.36 mg/L Ca as $CaCO_3$
    d. 254.72 mg/L Ca as $CaCO_3$

21. The saturation point of a chemical in water varies with _________.

    a. pH
    b. total dissolved solids (TDS)
    c. temperature
    d. All of the above

22. Why is it important to label the contents of all containers that carry materials or solutions in a water plant?

    a. To prevent mixing of incompatible materials
    b. To prevent unwanted reactions affecting health and safety
    c. To prevent losing a sample
    d. All of the above

23. The amount of oxygen that can remain dissolved in water is a function of the water's __________, and the __________ the water, the __________ the possible concentration of dissolved oxygen.

    a. pH, colder, lower
    b. pH, higher, higher
    c. temperature, colder, higher
    d. temperature, higher, higher

24. The precise location of sampling points depends on the __________ the distribution system.

    a. configuration of
    b. size and volume of
    c. pipe materials in
    d. contamination history within

25. Which is the most inexpensive pathogenic test?

    a. Viruses
    b. Coliforms
    c. *Giardia*
    d. *Legionella*

26. In order to test a laboratory and an analyst, the oversight agency gives the laboratory and analyst a __________ test.

    a. quality assurance
    b. quality control
    c. set of blind standards to
    d. performance evaluation

27. A virus would be considered a _________.

    a. dissolved solid
    b. colloidal solid
    c. biological solid
    d. Both b and c

28. Customer complaints of __________ are often caused by a __________.

    a. taste, odor, and color; home treatment device
    b. taste, odor, and color; water softener or water heater
    c. pink color on bathroom fixtures, water softener or home water treatment device
    d. pink color on bathroom fixtures, water heater

29. What are the additives in detergent that tie up hardness ions?

    a. Cationic and anionic fatty acids
    b. Fatty lipids
    c. Sequestering agents
    d. Calcium or magnesium salts of fatty acids

30. What rule requires the measurement of conductivity?

    a. Wastewater Reclamation Rule
    b. Arsenic Rule
    c. Lead and Copper Rule
    d. Stage 2 Disinfectants and Disinfection Byproducts Rule

31. Any good __________ household detergent is adequate for cleaning labware, and it is preferred that the detergent be of a __________ type.

    a. nonphosphate, nonliquid
    b. nonphosphate, liquid
    c. phosphate, nonliquid
    d. phosphate, liquid

32. Raw water temperature should be measured at a minimum of _________.

    a. every 4 hours
    b. every 8 hours
    c. every 12 hours
    d. every 24 hours

33. Find the amount of iron (Fe) removed per year from a plant that treats an average of 17.2 MGD, if the average raw water iron concentration is 0.46 mg/L and the removal efficiency is 82%. Round to the most appropriate answer.

    a. 19,000 lb/yr of Fe removed
    b. 20,000 lb/yr of Fe removed
    c. 21,000 lb/yr of Fe removed
    d. 23,200 lb/yr of Fe removed

34. If a treatment plant softens its water, where should samples be collected to determine whether the desired degree of softening has been achieved?

    a. Before the clearwell
    b. After the clearwell
    c. Immediately after softening and before the clearwell
    d. Immediately after softening and before the filters

35. As this parameter increases, corrosion decreases.

    a. Dissolved oxygen
    b. TDS
    c. Temperature
    d. Alkalinity

36. What threshold odor number (TON), also the USEPA secondary MCL, will begin to draw complaints from customers?

    a. 3 TON
    b. 4 TON
    c. 5 TON
    d. 6 TON

37. The turbidity reading from a meter at a water plant's clearwell is reading a zero immediately to the right of the decimal point, and the next three numbers to the right of the zero are 435. What should the operator record if three digits to the right of the decimal are needed?

    a. .043
    b. 0.043
    c. .044
    d. 0.044

38. Samples should not be collected from __________ or __________.

    a. chrome-plated, copper alloy pipe
    b. chrome-plated, CPVC
    c. sill cocks, faucets with hose threads
    d. sill cocks, most plastic materials

## Equipment Operation and Maintenance

1. The shaft's main function is to transmit _________ from the motor to the impeller.

   a. centrifugal force
   b. torque
   c. kinetic energy
   d. All of the above

2. The primary problem with hypochlorinators is that they can clog with __________. However, this material can be removed by pumping through the pump, hoses, and diffuser a dilute solution of__________.

   a. chloroorganic precipitates, HCl
   b. chloroorganic precipitates, $H_2SO_4$
   c. calcium carbonate, HCl
   d. calcium carbonate, $H_2SO_4$

3. Chlorine gas can kill a person after just a few breaths at concentrations as low as ________.

   a. 0.1%
   b. 0.2%
   c. 0.3%
   d. 0.5%

4. What is the the purpose of packing in a pump?

   a. keep oil or graphite on the shaft
   b. control water leakage along the pump's shaft
   c. reduce friction
   d. prevent wear on the shaft

5. What is motor horsepower?

   a. The portion of power delivered to a pump that is actually used to lift water
   b. The horsepower equivalent to the watts of electric power supplied to a motor
   c. The total power input to a motor and pump assembly
   d. The power supplied to a pump by a motor

6. What is the maximum withdrawal rate for a 1-ton chlorine container to prevent frosting if the backpressure is 35 psig and the container is at room temperature?

   a. 400 lb/d
   b. 450 lb/d
   c. 460 lb/d
   d. 500 lb/d

7. What is brake horsepower?

   a. The portion of power delivered to a pump that is actually used to lift water
   b. The horsepower equivalent to the watts of electric power supplied to a motor
   c. The total power input to a motor and pump assembly
   d. The power supplied to a pump by a motor

8. The primary reason filters are difficult to clean is because ________.
   a. sticky floc absorbs into the media
   b. sticky floc adsorbs to the media
   c. material collects on top of the media
   d. metal precipitates chemically react with the media
9. What is the water horsepower?
   a. The portion of power delivered to a pump that is actually used to lift water
   b. The horsepower equivalent to the watts of electric power supplied to a motor
   c. The total power input to a motor and pump assembly
   d. The power supplied to a pump by a motor
10. In fluoride saturator tanks, the chemical should be at least __________ deep at all times, and if the saturator treats more than 100 gpm, it should have a chemical depth of at least __________.
    a. 4-in., 6 in.
    b. 6-in., 10 in.
    c. 10-in., 12 in.
    d. 12-in., 18 in.
11. Chlorine scrubbers are designed to neutralize up to __________ of chlorine.
    a. 2,000 lb
    b. 4,000 lb
    c. 6,000 lb
    d. 8,000 lb
12. All of the following cause bearing failure, but which one is the major cause of bearing failure?
    a. Poor lubrication practices
    b. Contamination
    c. Vibration
    d. Faulty mounting
13. In a slow sand filter, once the sand layer is reduced to a depth of ______, new sand should be added to bring the filter bed back to the original depth.
    a. 1.0 ft
    b. 1.5 ft
    c. 2.0 ft
    d. 2.5 ft
14. What metal will be the anode from the following list because it is the most active?
    a. Tin
    b. Brass
    c. Mild steel
    d. Copper
15. A poor conductor of electricity would be ________.
    a. deionized water
    b. lead
    c. acid solutions
    d. zinc

16. Destruction of pathogenic organisms by chlorine is directly related to ________.

    a. chlorine concentration
    b. chlorine concentration, temperature, and pH
    c. chlorine concentration and contact time
    d. chlorine concentration and pH

17. The valves on chlorine cylinders and ton containers must comply with standards set by ________.

    a. Chemtrec
    b. Chlorine Institute
    c. USEPA
    d. the particular state's drinking water regulatory department

18. Packing replacement is done when ________.

    a. water leakage sprays out of the pump housing
    b. no further tightening can be done on the packing gland
    c. the packing gland bolts are exposed by more than 2½ in. above the nut
    d. Both a and b

19. Chlorine dioxide is __________ in the __________ and ___________.

    a. stable, presence of plastics, low temperatures
    b. stable, absence of light, elevated temperatures
    c. unstable, presence of concrete, corrosion inhibitors containing phosphates
    d. unstable, presence of high alkalinity, corrosion inhibitors containing phosphates

20. How often should screen equipment be inspected for corrosion?

    a. Weekly
    b. Monthly
    c. Quarterly
    d. Annually

21. Surface wash systems for filters agitate the top __________ of the media.

    a. 2 in.
    b. 4 in.
    c. 5 in.
    d. 6 in.

22. What disinfection treatment requires a monitoring procedure, other than residual monitoring, to determine its efficiency?

    a. Ozone
    b. Ultraviolet
    c. Chloramines
    d. Chlorine dioxide

23. As water flows through the volute, the pressure head ____________ and the velocity head ___________.

    a. increases, increases
    b. decreases, decreases
    c. decreases, increases
    d. increases, decreases

24. The best and most reliable method to find chlorine leaks is a chlorine detector. These detectors can sense leaks as small as __________.

    a. 1 ppm
    b. 2 ppm
    c. 3 ppm
    d. 5 ppm

25. The chlorine gas storage area should provide enough space for a ________.

    a. 10–30 day supply
    b. 20–45 day supply
    c. 30–60 day supply
    d. 70–100 day supply

26. A pressure-differential meter is a flow-measuring device that creates and measures a difference in pressure _________ to the rate of flow.

    a. proportional
    b. inversely proportional
    c. integral squared
    d. differential squared

27. Exhaust fan switches for chlorine feed rooms should be wired to __________ in the room.

    a. lights
    b. emergency lights
    c. the chlorine leak detector
    d. the chlorine alarm system

28. A bearing is selected according to its ________.

    a. load
    b. speed
    c. life expectancy
    d. All of the above

29. The following procedure is safe for changing out chlorine cylinders or 1-ton containers.

    a. After removing the brass outlet cap, clean the valve and threads with a wire brush.
    b. Always use a new washer.
    c. Always have a personal protective respirator nearby.
    d. All of the above

30. Patented backwashable pleated membrane filters have been designed specifically for the removal of ________.

    a. iron
    b. carbonates
    c. inorganics
    d. natural and synthetic organics

31. Where is the best place to install a vacuum breaker to prevent backflow on a fluoride system?

   a. Directly above the day tank
   b. As close to the chemical feeder as possible
   c. On the make-up water line
   d. Between the main storage tank and the day tank for fluoride

32. What is the most difficult and the most expensive physical measurement taken in a water plant?

   a. Turbidity
   b. Water flow
   c. Online total organic carbon analyzer
   d. Streaming current monitor

33. The main purpose of mechanical seals is to ________.

   a. keep lubrication in and dirt and other foreign materials out
   b. control water leakage
   c. keep lubricating grease in the seal
   d. hold the packing onto the shaft sleeve

34. The purpose of roller bearings is to ________.

   a. increase torque
   b. reduce friction
   c. support radial and axial loads
   d. Both b and c

35. When sodium fluoride solutions are used with hard water, insoluble compounds of __________ can form.

   a. iron fluoride
   b. iron and manganese fluoride
   c. calcium and magnesium carbonate
   d. calcium and magnesium fluoride

36. New packing rings should be staggered at ________.

   a. 45°
   b. 90°
   c. 120°
   d. 180°

37. What is the purpose of the graded gravel below the sand in a multimedia filter?

   a. To support the underdrain system
   b. To support the sand bed
   c. To prevent sand from entering the underdrain
   d. All of the above

38. Most fluoride dry feeders for large plants should have a vibrator to prevent __________.

   a. clogging
   b. free flowing
   c. bridging
   d. saturation

39. What disinfectant oxidant could have some THM formation?

    a. Potassium permanganate
    b. Ozone
    c. Chloramines
    d. Chlorine dioxide

40. Open impellers are usually used to pump water that contains ________.

    a. large solids
    b. any chemicals except acids
    c. alkaline chemicals only
    d. viscous polymers

41. What is the most appropriate chlorine gas solution controller to use if the chlorine demand rarely changes and it is necessary to compensate only for changes in the pumping rate?

    a. Residual flow control
    b. Compound loop control
    c. Flow proportional control
    d. Direct pressure control

42. Equipment's worst enemy(ies) is (are) ________.

    a. overheating
    b. dirt and moisture
    c. poor maintenance or none at all
    d. Both b and c

43. Chlorine gas has a pungent, noxious odor and a __________ color.

    a. greenish
    b. dark green
    c. greenish-yellow
    d. yellow

44. The purpose of the shaft's sleeves is to protect the shaft from ________.

    a. wear
    b. corrosion
    c. erosion
    d. All of the above

45. The bearings are protected from water leaking out of the stuffing box by the ________.

    a. shaft sleeve
    b. lantern ring
    c. slinger ring
    d. packing gland

46. What material lines the pipe that carries the corrosive chlorine solution from the injector to the point of application?

    a. Fiberglass
    b. PVC
    c. Stainless steel
    d. Aluminum

47. Pumping energy is measured in __________ head.

    a. elevation
    b. pressure
    c. velocity
    d. All of the above

48. Lime should be added to the water being treated _______.

    a. using a perforated pipe installed into the water being treated
    b. via several small-diameter jet sprays
    c. with as short a distance between the feeder and the water as possible
    d. with nonflexible pipe

49. What disinfectant oxidant is not at all affected by pH?

    a. Chlorine dioxide
    b. Chlorine
    c. Chloramines
    d. Ozone

50. What is the purpose of the chlorinator?

    a. Prevent blow off of chlorine gas
    b. Meter the chlorine gas safely and accurately
    c. Provide an ejector mechanism for the chlorine gas
    d. Provide a connection between the cylinder or container to the pipeline for application

51. Most pump packing is composed of ____________ with __________ lubrication.

    a. asbestos base, graphite
    b. flax, Teflon
    c. Teflon, inert oil
    d. aluminum or copper, inert oil

52. Most pressure-sensing devices used in water treatment plants rely on _______.

    a. hydraulic water pressure
    b. hydraulic oil pressure
    c. the metal-distortion principle
    d. mercury

53. What type of energy is produced when water moves through a pipe?

    a. Kinetic energy
    b. Mechanical energy
    c. Potential energy
    d. Velocity energy

54. Downflow fluoride saturator tanks with a gravel bed should be cleaned at least _________ depending on how fast hardness scale builds up.

    a. one to three times per year
    b. daily
    c. every two months
    d. every month

55. What is the major disadvantage of the manual solution feed method using sodium fluoride?
    a. It is not as accurate as other methods.
    b. The dust generated when added to the day tank
    c. Chemical needs to be manually weighed and added to a mixing tank
    d. All of the above

56. What is the brake horsepower (bhp) if 60 horsepower (hp) is supplied to a motor with 86% efficiency?
    a. 45 bhp
    b. 52 bhp
    c. 64 bhp
    d. 70 bhp

57. What is the motor horsepower (mhp), if 80 horsepower (hp = water horsepower or whp) is required to run a pump with a motor efficiency of 91% and a pump efficiency of 82%? Give answer to 2 significant figures.
    a. 59.7 mhp
    b. 60 mhp
    c. 107 mhp
    d. 110 mhp

58. Which type of impeller is best to use for pumping water with low volumes of solids?
    a. Closed
    b. Open
    c. Semi-open
    d. Split

59. Lubrication has the purpose(s) of _________.
    a. removing heat and reducing friction
    b. preventing corrosion
    c. reducing wear
    d. Doing all of the above

60. What is the only type of pump that can be operated against a closed valve?
    a. Vertical turbine pump
    b. Centrifugal pump
    c. Axial-flow pump
    d. Mixed-flow pump

## Source Water Characteristics

1. How is the gravel pack of a well disinfected?
    a. Sodium hypochlorite is added as gravel
    b. Calcium hypochlorite powder or tablets are added as gravel
    c. Chlorine gas bubbled into the well after gravel pack is installed
    d. Chlorine dioxide is added as gravel

2. __________ is most often found in __________ in rural areas, primarily as a result of __________.

   a. Nitrate, surface water, industrial processes
   b. Nitrate, shallow wells, agricultural contamination
   c. Nitrite, surface water, industrial processes
   d. Nitrite, shallow wells, agricultural contamination

3. What characteristic of water intensifies as water gets warmer?

   a. Taste and odor
   b. Pathogens
   c. Algae
   d. All of the above

4. Radon is a radioactive gas that is a natural decay product of _________.

   a. thorium
   b. neptunium
   c. radium
   d. potassium-argon

5. A water system that has radon at or below __________ standard does not have to treat its water.

   a. 300 pCi/L
   b. 400 pCi/L
   c. 500 pCi/L
   d. 600 pCi/L

6. Where will soft waters most likely occur?

   a. Where topsoil is thick
   b. Where limestone formations are present
   c. Where soil is sandy
   d. Where gypsum is present

7. What problems can excessive pumping rates cause on a groundwater system?

   a. Ground subsidence
   b. Seawater infiltration along coastal areas
   c. Excessive iron and manganese can be drawn into the well.
   d. All of the above

8. When zebra mussels die off, they decay and release large amounts of ________.

   a. calcium carbonate
   b. calcium-magnesium carbonate
   c. toxins
   d. phosphates

9. What is the normal pH range of groundwater?

   a. 6.5–8.0
   b. 6.0–8.0
   c. 6.0–7.5
   d. 6.5–7.5

10. What is the best protection for a well to prevent sediments from entering it after a period of rain?
    a. A properly installed working casing
    b. Properly sized gravel pack
    c. Surface grouting around well
    d. Properly sized well screen

11. What process will move water into the atmosphere?
    a. Transpiration
    b. Condensation
    c. Interception
    d. Infiltration

12. Water is considered to be too hard if it is greater than __________ as $CaCO_3$.
    a. 150 mg/L
    b. 200 mg/L
    c. 250 mg/L
    d. 300 mg/L

13. What is the primary reason that Asiatic clams are such a nuisance to operators?
    a. They are difficult to eliminate.
    b. They clog intake pipes.
    c. They clog treatment facilities.
    d. They cause tastes and odors.

14. What formation would be the least vulnerable to microbiological contamination?
    a. Siltstone
    b. Sandstone
    c. Limestone
    d. Fractured volcanic rock

15. What protozoan organism will cause eye infections and can be found in soil and water?
    a. *Toxoplasma*
    b. *Acanthamoeba*
    c. Microspore
    d. *Cyclospora*

16. All things being equal, which area would have the largest amount of runoff?
    a. Urban areas
    b. Frozen ground
    c. Steep slopes
    d. Clay soils

17. One method to measure the depth of water in a well is to use a __________ graduated in __________ of a foot.
    a. steel tape, hundredths
    b. steel tape, tenths and hundredths
    c. plastic tape marked with chalk, hundredths
    d. plastic tape marked with chalk, tenths and hundredths

18. Where may slimes come from at water treatment plants?
    a. Algae waste products
    b. Slime-producing diatomaceous algae
    c. Decaying organics
    d. Slime-producing bacteria and algae
19. Iron and manganese are found in groundwater when rock strata rich in these elements are subject to _________.
    a. high alkaline water that contains iron bacteria
    b. acidic water devoid of oxygen due to anaerobic activity
    c. organic acids and hydrogen sulfide gas
    d. acidic water that contains radon gas
20. What is the purpose of abandoning a well after it is no longer useful?
    a. To conserve the aquifer
    b. To prevent groundwater contamination
    c. To eliminate a physical hazard
    d. All of the above
21. If a sanitary survey finds that a groundwater system has a high risk of fecal contamination, what log inactivation is required for viruses?
    a. 2.5 logs
    b. 3.0 logs
    c. 3.5 logs
    d. 4.0 logs
22. Most groundwater and surface water sources contain _________.
    a. protozoan pathogens
    b. contaminants
    c. moderate-to-high levels of synthetic organic chemicals
    d. moderate-to-high levels of humic and fulvic acids
23. What method for evaluating a well's performance has a relatively short test time and does not indicate aquifer performance as completely as the other methods used?
    a. Recovery
    b. Drawdown
    c. Specific capacity
    d. Well yield
24. If rainwater flows over a surface that is exposed to decomposing organic matter in the soil, the _________ level of the water is increased, which will form _________ and thus increase the _________ of the water.
    a. nitrate, nitrites, pH
    b. carbon dioxide, carbonic acid, corrosivity
    c. organic, trihalomethanes, contamination
    d. hydrogen sulfide, a dilute sulfuric acid, corrosivity

25. Where will hard waters most likely occur?

    a. Where there is a thin topsoil
    b. Where sandy soil is present
    c. Where gypsum is present
    d. Where it rains a lot

26. An HAA5 compound would include __________.

    a. halonitriles
    b. haloaldehydes
    c. dibromoacetic acids
    d. chlorophenols

27. Deep groundwater aquifers tend to be of __________ quality, and most deep aquifers are __________ and __________.

    a. poorer, confined, do not recharge significantly locally
    b. poorer, unconfined, recharge very slowly
    c. high, unconfined, do not recharge significantly locally
    d. high, confined, recharge very slowly

28. What should a contingency plan for a wellhead protection plan encompass?

    a. Analytical and numerical models of water flow within the aquifer
    b. A sanitary survey of the watershed
    c. Detailed steps to take if contamination ever occurs
    d. Wellhead protection management

29. The total organic carbon in most waters is composed mainly of __________.

    a. lignins
    b. humic substances
    c. fulvic substances
    d. hemicellulose

30. What do zebra mussels typically attach themselves to?

    a. Intake structures
    b. Other organisms
    c. Slow-moving or stationary boats
    d. All of the above

31. What occurs when organic matter decomposes under anaerobic conditions?

    a. Iron begins to oxidize, and pH increases.
    b. Iron precipitates, greatly reducing algae growth.
    c. Hydrogen sulfide gas is produced, and pH increases.
    d. Hydrogen sulfide gas is produced, and pH decreases.

32. Manganese in groundwater is normally in the form of soluble __________.

    a. salts
    b. acids
    c. bases
    d. aldehydes

33. Creating an artificial wetland can improve water quality because many _________ that adhere to sediment are trapped by the artificial wetlands.

    a. algae species
    b. nutrients
    c. bacteria species
    d. bacteria and algae species

34. Water flowing downward into the earth is called __________.

    a. subsurface flow
    b. infiltration
    c. percolation
    d. gravity flow

35. The rate of reservoir __________ is generally a function of both the __________ in the watershed and how well the land is protected by __________ and management practices.

    a. erosion, number of trees, agricultural
    b. siltation, vegetation cover, agricultural
    c. erosion, type of soil, vegetation cover
    d. siltation, type of soil, vegetation cover

36. The first step in water source protection is __________.

    a. to map the watershed
    b. to determine the geographic area that needs protection
    c. to determine who exactly has an interest in protecting it
    d. Both b and c

37. Hydrology is the study of the __________ and circulation of water and its constituents as it moves through the atmosphere across the __________ surface and below the Earth's surface.

    a. chemistry, properties; Earth's
    b. distribution; land
    c. chemistry, properties; land
    d. properties, distribution; Earth's

38. Base flow would be water that flows from __________.

    a. surface waters to confined or unconfined aquifers
    b. groundwater springs and seeps to most streams
    c. mountainous area water to valleys
    d. the land to the sea

39. Highly mineralized water will have a __________ taste.

    a. fishy
    b. musty
    c. earthy
    d. medicinal

40. Which are disadvantages of surface intakes used on rivers or reservoirs?
    a. The water is warmer in the summer and may have surface ice in the winter.
    b. There is too much floating debris.
    c. The water quality is not usually as good as water at intermediate depths.
    d. All of the above.
41. Why is it important to maintain the lowest possible turbidity in water?
    a. Because turbidity can shield pathogenic microorganisms from the disinfectants
    b. The turbidity can combine chemically with the disinfectant, so there is less to kill or inactivate pathogens.
    c. It can, over time in the distribution system, produce tastes and odors.
    d. All of the above
42. In the drawdown method to evaluate a well's performance, the production well is pumped and water levels are periodically measured in __________.
    a. the production well and one or more monitoring wells
    b. an observation well
    c. two or more observation wells
    d. the production well and an observation well
43. From the following list, what waterborne pathogens will form cysts and survive for long periods of time despite even adverse environmental conditions?
    a. Some viruses and bacteria
    b. Some viruses and protozoa
    c. Some bacteria and protozoa
    d. Some bacteria and fungi
44. What type of algae blooms have been known to kill fish in ponds due to oxygen depletion and also causes livestock poisoning?
    a. Diatoms
    b. Flagellates
    c. Slime-producing algae
    d. Blue-green algae
45. If a well is relatively deep, properly constructed, and provides protection out to a fixed radius, it is probably adequate. However, what contamination can still cause problems?
    a. Chemicals
    b. *Cryptosporidium* oocysts
    c. Pathogenic bacteria
    d. Hepatitis A virus
46. What can cause well screens to plug?
    a. Bacteria
    b. Mechanical problems
    c. Chemical problems
    d. All of the above

47. If a river used as a water source was contaminated by an oil spill, what is the first thing that must be determined, and what is typically the simplest response to this problem?

    a. Determine length of time of the contamination episode; change to an alternative water source
    b. Determine the quantity of the spill; increase coagulant dose and treat the contaminated water accordingly
    c. Determine the length of time of the contamination episode; wait until the spill passes the intake before treating more water
    d. Determine the quantity of the spill; increase coagulant dose and treat the contaminated water accordingly

48. Most organizations say that soft water is less than __________ as $CaCO_3$.

    a. 60 mg/L
    b. 75 mg/L
    c. 100 mg/L
    d. 200 mg/L

49. What is the first barrier against pathogen intrusion?

    a. Watershed management
    b. Filtration
    c. Coagulation and flocculation
    d. Pre-disinfection or oxidation

50. Water in shallow wells will, on occasion, have relatively higher levels of __________ substances, whereas water from deep wells usually has relatively low levels of __________.

    a. synthetic, carbonates
    b. humic, organics
    c. synthetic, inorganics
    d. organic, inorganics

51. Springs are typically found __________.

    a. along faults
    b. at the base of hills
    c. on alluvial fans
    d. in the mountains

52. What is the purpose of grout placed around the well screen?

    a. To prevent soil from caving into the well
    b. To prevent water from traveling along the outside of the casing
    c. To support the casing
    d. Both b and c

53. Base flow is defined as __________ contributing water to a stream.

    a. surface runoff
    b. a spring
    c. groundwater
    d. a synthetic structure

54. Both types of rotary drilling methods can drill holes up to __________ in diameter, and both these types of drilling are __________.

    a. 4 ft, usually faster than the California method
    b. 5 ft, usually faster than the cable-tool drilling method
    c. 4 ft, more cost effective than the down-the-hole hammer method
    d. 5 ft, more cost effective than the cable-tool drilling method

55. The water level drop between the nonpumping level and the pumping level is called the _________.

    a. cone of depression
    b. radius of influence
    c. drawdown
    d. discharge level

## Security, Safety, Compliance, and Administrative Procedures

1. The simplest measuring and transmitting device is _________.

    a. float type
    b. bubbler units
    c. pressure activated
    d. sonic type

2. A confined space is defined as _________.

    a. a space that must have a limited means of entry and exit
    b. a space that must not be designed for continuous human occupancy
    c. a space that must be of a size and configuration that humans can enter and perform work
    d. All of the above

3. What organization approves self-contained breathing apparatuses (SCBAs)?

    a. ANSI
    b. NIOSH
    c. OSHA
    d. USEPA

4. An operator needs to pour some dilute acid into some water; what should be used for eye protection?

    a. Eye protection is not needed for this procedure
    b. Safety goggles
    c. Regular safety glasses
    d. Prescription safety glasses

5. Under the American National Standards Institute guidelines, as a general rule, what is considered as a last resort to safety?

    a. Engineering changes
    b. Administration changes
    c. Supervisor changes
    d. Personal protective equipment

6. The compliance for chlorite is based on the __________ of __________ chlorite monitoring locations each __________.

   a. average, three, month
   b. average, five, quarter
   c. running annual average, three, month
   d. running annual average, five, quarter

7. Radon is a gas most often found in __________.

   a. metamorphic rock formations
   b. sedimentary clays and silts
   c. granite formations
   d. volcanic rocks

8. The Interim Enhanced Surface Water Treatment Rule requires that the water industry provide a minimum __________ of __________.

   a. 2-log removal, *Cryptosporidium*
   b. 3-log removal, *Cryptosporidium*
   c. 2-log removal, *Giardia*
   d. 3-log removal, *Giardia*

9. What gas will cause olfactory fatigue?

   a. Radon
   b. Methane
   c. Hydrogen sulfide
   d. Nitrous oxide

10. What is the main cause of THMs when chlorination is used?

    a. Algae
    b. Bacteria
    c. Humic and fulvic acids
    d. Proteins

11. Eyewashes should be located __________.

    a. in every pump room
    b. within 25 ft of all chemical storage areas
    c. in every laboratory
    d. within 10 ft of every chemical storage tank

12. The original SDWA and the 1986 amendments focused on what aspect to provide safe drinking water?

    a. Treatment
    b. Water source water protection
    c. Water quality out of the treatment plant and in the distribution system
    d. Funding for water systems improvements

13. If phosphates are used for corrosion control, mains should be flushed regularly with phosphate results recorded __________.

    a. in 5-, 10-, and 15-min intervals
    b. in 10-min intervals for 1 hour
    c. in 15-min intervals for 1 hour
    d. every hour for 4 hours

14. What chemical form, dry or liquid, typically has the advantage over the other in the following categories?

    1. Simplified chemical storage
    2. Reduced safety hazards
    3. Less space to store
    4. Easier to handle
    5. Will not cake
    6. Easier to feed
    7. Less dirt and mess

    a. Dry: 3, 4, and 6; liquid: 1, 2, 5, and 7
    b. Dry: 1, 2, and 3; liquid: 4–7
    c. Dry: 1 and 2; liquid: 3–7
    d. Dry: none; liquid: 1–7

15. Systems that use chlorine dioxide will have to monitor for __________.

    a. chlorite
    b. chlorate
    c. bromate
    d. bromide

16. If a boil order is issued by a water utility, customers should boil the water for at least _________ or longer if at high altitude.

    a. 3 min
    b. 5 min
    c. 10 min
    d. 12 min

17. What regulation specifies the percentage removal of TOC?

    a. Interim Enhanced Surface Water Treatment Rule
    b. Surface Water Treatment Rule
    c. Stage 1 Disinfectants and Disinfection Byproducts Rule
    d. Stage 2 Disinfectants and Disinfection Byproducts Rule

18. What percentage of deaths in confined spaces are the deaths of would-be rescuers?

    a. 25%
    b. 50%
    c. 67%
    d. 75%

19. What emergency repair kit is used for a chlorine cylinder?

    a. Kit A
    b. Kit B
    c. Kit C
    d. Kit D

20. What organization provides testing for point-of-use or point-of-entry devices?

    a. County public health departments
    b. State public health departments
    c. NSF International
    d. Water Quality Association

21. What water use system regulates water use so that overdraft cannot occur?

    a. Appropriative rights
    b. Appropriation-permit system
    c. Riparian doctrine
    d. Correlative rights

22. In the Total Coliform Rule, the USEPA has set the health goal for total coliforms at __________.

    a. zero
    b. one positive per 40
    c. 5% of samples
    d. <5% of samples

23. The Long-Term 1 Enhanced Surface Water Treatment Rule applies to surface water systems and groundwater under the direct influence of surface water serving fewer than __________.

    a. 3,300
    b. 5,000
    c. 10,000
    d. 25,000

24. What type of process control is used when the process needs a wide proportional band for stable operation?

    a. Floating proportional control
    b. Proportional control
    c. Proportional plus reset control
    d. Proportional plus reset plus derivative control

25. What log removal is required for viruses under the Groundwater Rule?

    a. 2.5 logs
    b. 3 logs
    c. 3.5 logs
    d. 4 logs

26. The Filter Backwash Recycle Rule is particularly associated with __________.

    a. viruses
    b. *Cryptosporidium*
    c. bacterial pathogens
    d. coliforms

27. In the Stage 1 Disinfectants and Disinfection Byproducts Rule, the maximum chlorine dioxide level is set at __________.

    a. 0.8 mg/L
    b. 1.0 mg/L
    c. 1.2 mg/L
    d. 1.5 mg/L

28. There is not a maximum contaminant level for total dissolved solids (TDS), but there is a secondary standard for TDS of __________. TDS is a gross measure of the __________ of the water because __________ substances are usually a minor part of the total.

    a. 250 mg/L, organics, volatile
    b. 500 mg/L, organics, volatile
    c. 250 mg/L, inorganics, organic
    d. 500 mg/L, inorganics, organic

29. _________ is the organization responsible for testing and approving water treatment chemicals and components.

    a. AWWA
    b. NSF International
    c. Association of State Drinking Water Administrators
    d. Conference of State Health and Environmental Managers

30. What type of process control will not lose control action when the signal is lost?

    a. Feedback control
    b. Proportional reset control
    c. Floating control
    d. Floating proportional control

31. *Cryptosporidium* is very susceptible to __________.

    a. chlorine dioxide
    b. chlorine
    c. ultraviolet light
    d. ozone

32. What parameter has an MCL?

    a. Ammonia
    b. Lead
    c. Phosphate
    d. Nitrite

33. Tasting a little bit of a laboratory reagent is __________.

    a. permissible for bases only
    b. permissible for acids only
    c. permissible as long as there is no skull and crossbones on the label
    d. dangerous and can even be fatal

34. Operators that use a self-contained breathing apparatus need a medical review every __________. Which organization requires fit testing of tight-fitting respirators?

   a. year, OSHA
   b. year, state public health departments
   c. two years, OSHA
   d. two years, state public health departments

35. What is the primary reason for developing and maintaining safe working conditions and practices?

   a. To eliminate lost work time as much as possible
   b. To eliminate injuries
   c. To eliminate or reduce medical costs
   d. To reduce costs and litigation for the utility

36. Water systems that fail to meet the lead and copper action levels required by the Lead and Copper Rule must take corrective action as directed by the __________.

   a. USEPA
   b. county regulatory agency
   c. state regulatory agency
   d. utility's management

37. Exposure over many years to drinking water containing copper above the action level could increase the risk of __________.

   a. liver and kidney damage
   b. heart failure
   c. thyroid failure
   d. intestinal cancer

38. If a filter's turbidity exceeds __________ ntu in the first 4 hours of a given filter run for two consecutive __________ min samples, the results must be reported within __________ days following the month of occurrence.

   a. 0.3, 5-, 5
   b. 0.3, 15-, 10
   c. 0.5, 15-, 10
   d. 1.0, 30-, 14

39. Who is responsible for knowing the hazards of a confined space?

   a. Entry supervisor
   b. Authorized attendant
   c. Authorized entrant
   d. All of the above

40. Laboratories should be equipped with __________ and __________.

   a. Class A, Class B fire extinguishers
   b. Class B, Class C fire extinguishers
   c. all-purpose fire extinguishers, fire blankets
   d. all-purpose fire extinguishers, a burn kit

41. If a system changes from free chlorine to chloramines, hospitals and kidney dialysis centers must be alerted because it can cause __________.

    a. hyperanaemia
    b. hyperhomocysteinaemia
    c. hemolytic anemia
    d. hypochromic anemia

42. The locational running annual average applies to what USEPA regulation?

    a. Interim Enhanced Surface Water Treatment Rule
    b. Ground Water Rule
    c. Stage 1 Disinfectants and Disinfection Byproducts Rule
    d. Stage 2 Disinfectants and Disinfection Byproducts Rule

43. The Revised Total Coliform Rule eliminated the MCLG and MCL for __________, and established an MCLG and MCL of zero for __________.

    a. *Giardia,* total coliforms
    b. *Cryptosporidium,* total coliforms
    c. *Giardia, E. coli*
    d. total coliforms, *E. coli*

44. No antibiotic treatment currently exists for __________.

    a. *Salmonella*
    b. *Shigella*
    c. *Cryptosporidium*
    d. *Entamoeba*

45. What radioactive contaminant will most probably cause lung cancer?

    a. Radium
    b. Uranium
    c. Radon
    d. Plutonium

46. The maximum contaminant level (MCL) is the allowable concentration of a contaminant in drinking water. Who establishes the MCLs?

    a. USEPA and AWWA
    b. State and federal regulations
    c. AWWA and local public health departments
    d. NSF/ANSI

47. Who is responsible for maintaining safe working conditions, the physical condition of their facilities, and supporting policies that encourage the safe performance of work duties?

    a. Utility manager
    b. Supervisor
    c. Chief operator
    d. Every individual associated with the water utility

48. Hydrogen sulfide is a poisonous gas and is colorless, __________ and __________ than air.

    a. inert, heavier
    b. explosive, lighter
    c. explosive, heavier
    d. inert, lighter

49. To comply with the enhanced coagulation requirements, a softening plant would need to remove __________ of __________ hardness.

    a. 5 mg/L, carbonate
    b. 10 mg/L, carbonate
    c. 5 mg/L, magnesium
    d. 10 mg/L, magnesium

50. Just before an operator enters a non permit confined space, what should have been done first?

    a. The paperwork should have been filled out near the confined space.
    b. The gas monitor should have been calibrated or bump tested with a calibration within the last 30 days.
    c. The operator should have checked the atmosphere of the confined space to declassify it as a non permit space.
    d. All of the above

51. In the Stage 1 Disinfectants and Disinfection Byproducts Rule, the maximum level of chloramines is set at __________.

    a. 3.0 mg/L
    b. 4.0 mg/L
    c. 5.0 mg/L
    d. 6.0 mg/L

52. The color of copper staining of fixtures is usually __________.

    a. greenish red
    b. blue green
    c. pinkish red
    d. reddish brown

53. The combined filter effluent turbidity must be __________ ntu for __________ % of samples collected each month; and none of the samples can be __________ ntu.

    a. =0.3, 95, >1
    b. <0.3, 98, =1
    c. <0.3, 98, =0.5
    d. =0.3, 95, >1

54. A positive total coliform result requires three repeat samples, one at the site of the positive total coliform site and two others within __________ service taps upstream and downstream of the positive site.

    a. two
    b. three
    c. four
    d. five

55. In the Long-Term 1 Enhanced Surface Water Treatment Rule, water systems are required to achieve a __________ removal of __________.

    a. 2-log, *Giardia* cysts
    b. 2-log, *Cryptosporidium*
    c. 3-log, *Giardia* cysts
    d. 3-log, *Cryptosporidium*

56. Who is primarily responsible for safety at a water plant?

    a. Utility supervisor
    b. Supervisor
    c. Chief operator
    d. Every individual associated with the water utility

57. What USEPA rule may require a system to provide a special customer education program?

    a. Total Coliform Rule
    b. Surface Water Treatment Rule
    c. Stage 2 Disinfectants and Disinfection Byproducts Rule
    d. Lead and Copper Rule

58. What gas can come off storage batteries, requiring batteries be in well-ventilated areas?

    a. Carbon monoxide
    b. Carbon dioxide
    c. Hydrogen
    d. Hydrogen sulfide

59. How does fluoride adversely affect the human body besides possible mottled teeth?

    a. Causes bone disease
    b. Causes heart damage
    c. Causes liver disease
    d. Causes kidney damage

60. What is the maximum residual disinfectant level for chlorine?

    a. 4.0 mg/L free chlorine
    b. 4.0 mg/L total chlorine
    c. 5.0 mg/L free chlorine
    d. 5.0 mg/L total chlorine

61. The 1986 Safe Drinking Water Act set up a timetable under which the USEPA was required to develop primary standards for __________ contaminants among other provisions, including __________.

    a. 63, a requirement to assess and possibly regulate 10 new contaminants every 5 years
    b. 83, prohibiting use of lead products to convey drinking water
    c. 63, a specific MCLG for carcinogens
    d. 83, lowering the arsenic MCL

62. Safety showers should _________.

    a. always be turned off manually, never by themselves
    b. flow a torrent of water in a uniform pattern
    c. be easy to reach and turn on
    d. All of the above

63. Usually, pathogen exposure presents _________ rather than __________ risk, and if pathogens are detected, it will result in __________.

    a. acute, chronic, boil-water notices
    b. acute, chronic, steep fines on the utility
    c. chronic, acute, boil-water notices
    d. chronic, acute, steep fines on the utility

64. Noncarcinogenic organic chemical effects are usually on the ________.

 a. central nervous system
 b. pancreas
 c. gastrointestinal system
 d. small intestines

65. What is a disadvantage of biological treatment?

 a. It often causes tastes and odors.
 b. The risk of introducing pathogenic organisms
 c. Higher chlorine demand
 d. Increased microbial growth in the distribution system

66. If soda ash dust gets into the eyes, they should be flushed with warm water for at least ________.

 a. 10 min
 b. 15 min
 c. 20 min
 d. 30 min

67. Ingesting drinking water containing too much copper can lead to ________.

 a. Lesch Nyhan Syndrome
 b. pseudomyxoma peritonei
 c. Wilson's disease
 d. Menkes disease

68. The burning of coal will produce _________. A process known as _________ converts the sulfur-containing molecule into sulfuric acid.

 a. sulfur oxide, UV oxidation
 b. sulfur dioxide, photochemical oxidation
 c. sulfur oxide, photochemical oxidation
 d. sulfur dioxide, UV oxidation

69. Hydrogen sulfide is a very dangerous gas because at high concentrations it ________.

 a. disrupts the nervous system
 b. paralyzes one's sense of smell
 c. kills in about 1 hour
 d. destroys one's ability to think clearly

70. In drinking water, the coliform group of bacteria is used as an indicator organism for the presence of fecal contamination, which may indicate the presence of pathogens. The criteria for an ideal organism for pathogens in drinking water should be that they ________.

 a. are present at least 98% of the time when coliforms are present
 b. are absent more than 98% of the time when fecal contamination is absent
 c. survive about as long as pathogens in the water
 d. are easily identified

71. Currently, what is the principal use of reclaimed water?

 a. Golf courses and lawn strips along streets
 b. Agricultural
 c. Flushing toilets
 d. Domestic lawn watering

72. The Safe Drinking Water Act of 1990 requires water systems to _________.

    a. disinfect the water
    b. disinfect and filter the water
    c. implement conventional treatment processes
    d. have a 4-log removal of *Giardia* and *Cryptosporidium*

73. What is the unit of gamma radiation?

    a. Coulombs
    b. Neutrinos
    c. Rem
    d. Ergs

74. If variances and waivers are not granted after initial sampling, a public water system should sample for Phase I, II, and V contaminants every __________.

    a. year
    b. 2 years
    c. 3 years
    d. 5 years

75. Ammonia may be fatal at concentrations of ________.

    a. 500 ppm
    b. 1,000 ppm
    c. 1,500 ppm
    d. 2,000 ppm

76. Records need to be reported for any testing required for filter exemptions under the ________.

    a. Surface Water Treatment Rule (SWTR)
    b. Groundwater Rule
    c. Interim Enhanced SWTR
    d. Long-Term 1 or 2 Enhanced SWTR

77. What is the main focus of the Groundwater Rule?

    a. Synthetic organic contamination
    b. Nitrite and nitrate contamination
    c. Inorganic contamination
    d. Fecal contamination

78. Water treatment facilities should keep a 30-day supply of chemicals because __________, but not more than a 60-day supply because __________.

    a. natural or man-made disasters might occur, of storage issues
    b. delivery problems may occur, the chemicals over time lose their effectiveness
    c. natural or man-made disasters might occur, the chemicals over time lose their effectiveness
    d. delivery problems may occur, of storage issues

# II. Water Treatment, Advanced Questions



## Treatment Process

1. If clarification and filtration follow aeration, what process will most probably suffer most if too much aeration occurs?

   a. Sedimentation
   b. Coagulation
   c. Filtration
   d. Post-disinfection

2. Jar tests are used to determine the powdered activated carbon dose. The glassware used in these jar tests need to be cleaned with __________ and rinsed with __________ water.

   a. phosphate-free detergent, deionized
   b. phosphate-free detergent, copious amounts of
   c. unscented detergent, odor-free
   d. unscented detergent, copious amounts of

3. The rate of formation of __________ increases with increasing pH.

   a. HAA5
   b. nitrates
   c. THMs
   d. organophosphates

4. In variable declining-rate filtration, what is placed in the effluent line of each filter to prevent excessive flow rates?

   a. A rate-of-flow controller
   b. A flow-restricting orifice plate
   c. A variable solenoid-controlled butterfly valve
   d. Pressure-controller valve

5. A tracer study uses a substance that can readily be identified in water to determine the __________ in a basin, pipe, or channel.

   a. reaction time and flow pattern
   b. reaction time and distribution
   c. flow pattern and distribution
   d. distribution and detention time

6. Both the amount and types of disinfection byproducts formed depend not only on the organic precursors present but also on the _________.
   a. type of disinfectant used and contact time with that disinfectant
   b. alkalinity
   c. temperature
   d. conductivity

7. What is the best practice to follow for a filter that is air-bound?
   a. Remove from service and backwash thoroughly
   b. Close the effluent valve and fill the filter as high as permissible. The extra pressure on the media will force the air out slowly. This takes about 15–45 min for most filters. After bubbles stop rising from the filter for 5 min, the filter is ready to be put back in service.
   c. Simply bump the filter to release all the gases. It may need to be bumped several times until bubbles stop being released.
   d. Both a and b

8. In manganese greensand filters, the greensand is usually capped with at least _________.
   a. 6-in. of anthracite
   b. 6-in. of coarse sand and 6 in. of anthracite
   c. 12-in. of garnet
   d. 6-in. of coarse sand and 12 in. of anthracite

9. Lamella plates work by _________.
   a. using plates that are inclined by 40°
   b. first using plates inclined downward by 20° and then upward into plates inclined by 45°
   c. using plates directed downward at 45° then upward at 45°
   d. using parallel plates that direct the water downward then upward at 55°

10. The flow through a sedimentation basin should not exceed ________.
    a. 0.5 ft/min
    b. 0.8 ft/min
    c. 1.0 ft/min
    d. 1.5 ft/min

11. What is the mechanism by which potassium permanganate works when used to control algae?
    a. It disrupts the algae's DNA, thus killing it.
    b. It precipitates iron, which is needed by the algae for chlorophyll production.
    c. It reacts with the algae cell wall, causing it to lyse and thus die.
    d. It disrupts the algae's mitochondria, thus killing it.

12. If THMs are still a problem after softening, and changing the disinfectant is not an option, what other treatment process can be used and what is its disadvantage?
    a. Granular activated carbon, rather expensive
    b. Powdered carbon, high maintenance
    c. Tube settlers, difficult to clean
    d. Coagulant aid, additional maintenance

13. What process is currently used for softening brackish water?

    a. Nanofiltration
    b. Electrodialysis
    c. Freezing
    d. Lime–soda ash process

14. Greensand filters are often capped to enhance filter performance and improve backwash efficiency. It is common to cap them with ________.

    a. 1–2 in. of granular activated carbon (GAC)
    b. 12-in. of GAC
    c. 6-in. of anthracite
    d. 18-in. of anthracite

15. What other effect does aeration have besides oxidizing iron and manganese?

    a. It removes organics via reduction.
    b. It removes $CO_2$, which enhances oxidation because it raises the pH.
    c. It inactivates *Cryptosporidium*, including their oocysts.
    d. It destroys or inactivates all bacterial pathogens.

16. Chloramines will be used when tastes and odors are a problem, particularly when the source water contains ________.

    a. humic acids
    b. phenols
    c. tannic acids
    d. capsaicin

17. What treatment chemical is corrosive, dense, and an oily liquid and is available in strengths of 62%, 78%, and 93%?

    a. Sodium hexametaphosphate
    b. Zinc orthophosphate
    c. Caustic soda
    d. Sulfuric acid

18. Besides oxidizing iron and manganese, potassium permanganate has been found to be most effective in controlling ________.

    a. zebra mussels
    b. *Vibrio cholerae*
    c. *Salmonella typhi*
    d. protozoans

19. Dissolved air flotation has achieved a ________ removal of *Cryptosporidium*, which compares to a removal rate of less than 0.5 log using conventional sedimentation.

    a. 1.0-log
    b. 1.5-log
    c. 2.0-log
    d. 2.5-log

20. _________ will outperform _________ in open filters in the removal of organics.

 a. Anthracite, GAC
 b. GAC, anthracite
 c. PAC, slow sand filters
 d. PAC, GAC

21. Iron complexes with _________ are _________ in chemistry, and they are not removed by _________.

 a. sulfate, cationic, cation exchange
 b. sulfate, anionic, anion exchange
 c. organic matter, cationic, mixed bed resins
 d. organic matter, anionic, cation exchange

22. A special problem with granular activated carbon filters is the rapid growth of _________.

 a. helminths
 b. amoebas
 c. algae
 d. bacteria

23. How is aluminum oxide activated?

 a. Oxygen and heat
 b. Sulfuric acid and heat
 c. Heat and washing with sodium hydroxide
 d. Carbon dioxide wash with HCl and heat

24. What is the most common scale-forming precipitate?

 a. $MgCO_3$
 b. $MgCl_2$
 c. $CaCO_3$
 d. $CaSO_4$

25. What membrane configurations are currently available and used in drinking water treatment plants?

 a. Spiral wound and hollow fiber
 b. Spiral wound and tubular fiber
 c. Tubular and hollow fiber
 d. Hollow fiber and plate and frame

26. The _________ and settling of calcium carbonate and magnesium hydroxide precipitates occur more _________ in the presence of _________.

 a. flocculation, quickly, previously formed floc particles
 b. softening reactions, quickly, calcium carbonate granules
 c. coagulation, slowly, calcium carbonate granules
 d. chemical reactions, quickly, too much sludge so frequent removal is necessary

27. A low-cost method of feeding powdered activated carbon (PAC) is to use a _________, where precise feed is _________.

 a. helix-type feeder, primary
 b. helix-type feeder, secondary
 c. slurry feed system, primary
 d. slurry feed system, secondary

28. If powered activated carbon (PAC) does not remove tastes and odors, it could be that it is added too close to the chlorine feed. To solve this problem, PAC should be added so that it has at least __________ of contact time before chlorine is added.

a. 10 min
b. 15 min
c. 20 min
d. 25 min

29. To prevent serious loss of media, it is recommended that the surface washers be turned off __________ min before the end of the main backwash.

a. 1–2
b. 3–5
c. 4–7
d. 7–10

30. The lime–soda ash process has the following characteristics:

a. It softens water at a moderate cost.
b. It can be used for both groundwater and surface water.
c. It is most often used at small treatment plants.
d. All of the above

31. No preliminary contact or sedimentation chambers are necessary for _________.

a. aeration with oxygen
b. potassium permanganate oxidation
c. manganese greensand filters
d. ozone oxidation

32. What is the percentage of copper in copper sulfate ($CuSO_4$)?

a. 39.80% Cu
b. 39.81% Cu
c. 39.84% Cu
d. 39.88% Cu

33. Diatomaceous earth filtration is used for raw water that has a maximum turbidity of ________.

a. 5 ntu
b. 10 ntu
c. 20 ntu
d. 40 ntu

34. What two methods are commonly used for drinking water supplies to control algae at the source?

a. Copper hydrate and activated carbon
b. Potassium permanganate and anthracite
c. Potassium sulfate and anthracite
d. Copper sulfate and powdered activated carbon

35. What could cause eddies in a clarifier or sedimentation basin?

    a. Uneven flow over the weir
    b. Broken baffle
    c. Sludge level too high
    d. Gusting winds

36. What chemical used for oxidizing iron and manganese needs a reaction basin?

    a. Chlorine
    b. Chlorine dioxide
    c. Ozone
    d. Potassium permanganate

37. What is produced after colloids have just become charge neutralized?

    a. Stabilized microfloc
    b. Stabilized colloid
    c. Macrofloc
    d. Agglomerated microfloc

38. In a dual media filter, it is best to have __________ backwash rates because each type of media has a different __________.

    a. two, particle size and thus surface area
    b. two, specific gravity
    c. three, particle size and thus surface area
    d. three, specific gravity

39. Reverse osmosis (RO) units also have cartridge filters installed immediately ahead of the RO unit to remove suspended matter down to ________.

    a. 1–3 mm
    b. 3–5 mm
    c. 5–10 mm
    d. 10–15 mm

40. What chemical used for oxidizing iron and manganese will react with hydrogen sulfide, cyanides, phenols, and other taste and odor compounds and will not form trihalomethanes?

    a. Chlorine dioxide
    b. Potassium permanganate
    c. Ozone
    d. Chlorine

41. How does bentonite clay improve coagulation?

    a. Adds alkalinity
    b. Increases pH
    c. Makes the floc heavier
    d. Stabilizes the charge on the particle

42. A filter core sample is best collected for solids retention analysis ________.

    a. when the filter media is dry
    b. when the filter media has been drained and the media is still damp
    c. when the water has been drained to just above the media
    d. at any of the above scenarios

43. If sludge removal scrapers are installed at a plant, it is not necessary to interrupt plant operations to take the basin out of service as frequently as basins without scrapers. However, the basins will still need to be dewatered and hose-blasted at least ________.

a. monthly
b. once per year
c. every 2 years
d. every 3 years

44. To maintain the optimum powdered activated carbon (PAC) dose, the threshold odor tests should be performed on the raw water and finished water while PAC is being fed at least ________.

a. daily
b. weekly
c. bimonthly
d. monthly

45. Engineers usually design the height of a filter's backwash collection trough to allow for __________ bed expansion without loss of media into the troughs.

a. 25%
b. 30%
c. 40%
d. 50%

46. What is the best way to save on chemical costs when softening hard water if the Surface Water Treatment Rule does not apply to the system?

a. Reverse osmosis
b. Enhanced coagulation
c. Lime–soda ash
d. Split treatment

47. Excess air from aeration will cause air to attach to floc, making the floc __________ and not settle. It may also attach to the filter media, if the water __________ as it passes through the filter media, which, if it continues long enough, will cause __________.

a. disperse, warms, air binding
b. disperse, cools, clogging
c. float, warms, air binding
d. float, cools, clogging

48. The pores in carbon are created by exposing the carbon to __________. This process is known as __________ the carbon.

a. very high heat in the presence of steam, carbonizing
b. high pressure steam, tunneling
c. high pressure steam, cracking
d. very high heat in the presence of steam, activating

49. Dewatering of alum and ferric sludge is difficult because the water is ________.

a. covalently bonded to the hydroxide ions in the floc
b. covalently bonded to the aluminum or ferric cations in the floc
c. chemically bound to the aluminum and ferric hydroxide floc
d. physically bound to the aluminum and ferric cations

50. Regarding the two-stage recarbonation process, the raw water pH is elevated to __________ or greater with lime. After that stage, carbon dioxide is added to lower the pH to __________.

    a. 10, 9.0–9.5
    b. 11, 10.0–10.6
    c. 10, 9.5–9.9
    d. 12, 10.4–10.7

51. What is an advantage that coagulants such as ferric sulfate have over alum?

    a. They are not as affected by temperature.
    b. They work over a wider pH range.
    c. They need very little alkalinity.
    d. They are not affected by prechlorination.

52. The __________ process is used in the water industry primarily for the removal of __________.

    a. adsorption, organic substances
    b. adsorption, inorganic substances
    c. absorption, organic substances
    d. absorption, inorganic substances

53. Water that is highly colored with organic material requires a __________ pH for coagulation; therefore, __________ before __________ may be necessary.

    a. lower, acid addition, flocculation
    b. higher, alkalinity addition, flocculation
    c. lower, acid addition, coagulation
    d. higher, alkalinity addition, coagulation

54. __________ is often added to provide __________ and facilitate iron precipitation.

    a. Lime, alkalinity
    b. Soda ash, a higher pH
    c. Caustic, basicity
    d. Sodium bicarbonate, a higher pH

55. A hazardous measuring device would most likely be __________.

    a. a pressure-activated level transmitter.
    b. a laser particle counter.
    c. a nephelometric meter.
    d. a manometer.

56. What acid is even more corrosive when diluted than when it is pure?

    a. Sulfuric acid
    b. Hydrochloric acid
    c. Nitric acid
    d. Phosphoric acid

57. What gas(es) does electrolytic flotation use?

    a. Air
    b. Nitrogen
    c. Carbon dioxide
    d. Hydrogen and oxygen

58. Reverse osmosis membranes need cleaning if the pressure drop through the unit increases by _________.

   a. 10%
   b. 15%
   c. 20%
   d. 25%

59. What alum sludge dewatering process can attain at least 20% solids by weight?

   a. Drying lagoon
   b. Vacuum filters
   c. Sand drying beds
   d. Freezing and thawing method

60. __________ are used to regenerate ion-exchange systems and are also used to __________ the pH of water treated by nanofiltration and __________.

   a. Bases, raise, microfiltration
   b. Bases, raise, reverse osmosis membranes
   c. Acids, lower, reverse osmosis membranes
   d. Acids, lower, microfiltration

61. Water flowing through a full pipeline has a velocity of 1.75 ft/s. If the flow through the pipe is 0.35 $ft^3/s$, what is the diameter in inches of the pipeline?

   a. 3.0 in.
   b. 4.0 in.
   c. 5.0 in.
   d. 6.0 in.

62. What is the velocity of flow in ft/s for an 8.0-in.-diameter pipe if it delivers 182 gpm?

   a. 0.9 ft/s
   b. 1.2 ft/s
   c. 3.6 ft/s
   d. 8.7 ft/s

63. The iron in a corroding pipe releases two electrons at the anode. What will these electrons combine with in the pipe?

   a. Metal cations in the flowing water at the cathode
   b. Organics in the flowing water
   c. Ferrous hydroxide at the cathode
   d. Two hydrogen ($H^+$) ions to form hydrogen gas

64. An organic contaminant from coagulants would most likely include _________.

   a. epichlorohydrin
   b. polynuclear aromatic hydrocarbons
   c. polychlorinated biphenyls
   d. tetrachloroethylene

65. What pH is needed to form the magnesium hydroxide precipitate?

   a. Above 9.0
   b. Above 9.2
   c. Above 9.5
   d. Above 10.6

66. Chlorite monitoring requirements include __________ sampling and analysis at the distribution system's entry point and __________ sampling and analysis at locations in the distribution system.

    a. daily, quarterly
    b. daily, monthly
    c. monthly, quarterly
    d. monthly, monthly

67. In dual media, if the ratio of anthracite size to sand size is close to __________, a sharp interface will be created and __________ and possibly _________ will result.

    a. 2, too much intermixing, poorer quality filtrate
    b. 2, excessive head losses, mudball formation
    c. 4, excessive head losses, mudball formation
    d. 4, too much intermixing, poorer quality filtrate

68. What type of aerators have riffle plates?

    a. Cone aerators
    b. Slat-and-coke tray aerators
    c. Cascade aerators
    d. Spray aerators

69. What is the primary reason for filters to have rate-of-flow controllers on the filtered water effluent?

    a. Prevent bumping
    b. Prevent cracks from forming in the filter
    c. Prevent flow surges
    d. Prevent underdrain damage

70. Scale is formed from the combination of ________.

    a. monovalent metallic ions with carbonates or sulfates
    b. monovalent metallic cations and minerals dissolved in water
    c. divalent metallic ions with carbonates or sulfates
    d. divalent metallic cations and minerals dissolved in water

71. A plant uses a coagulant aid that weighs 10.05 lb/gal to help treat 10.8 mgd. The coagulant aid's strength is 50%. The results of a drawdown test are 112 mL for the coagulant aid in 5 min. What is the coagulant aid dosage in ppb?

    a. 470 ppb of coagulant aid
    b. 475 ppb of coagulant aid
    c. 480 ppb of coagulant aid
    d. 485 ppb of coagulant aid

72. Water treatment plants that use powdered activated carbon consistently have found the handling problems are less by using ________.

    a. dry gravimetric feeders
    b. volumetric feeders
    c. slurry feed system
    d. helix-type feeders

73. A 0.5*N* solution of NaOH is to be prepared. If 1 L of solution is desired, how many grams of NaOH are required? The gram formula for NaOH is 40.00.

a. 10 g of NaOH dissolved in 1 L of deionized water
b. 20 g of NaOH dissolved in 1 L of deionized water
c. 40 g of NaOH dissolved in 1 L of deionized water
d. 80 g of NaOH dissolved in 1 L of deionized water

74. Alum sludge that has been dewatered in a lagoon for a year may still contain more than ________ of the original water.

a. 75%
b. 80%
c. 90%
d. 95%

75. What membrane process will remove hardness?

a. Nanofiltration and reverse osmosis
b. Nanofiltration and ultrafiltration
c. Ultrafiltration and microfiltration
d. Reverse osmosis and ultrafiltration

76. What corrosion prevention chemical may actually cause tuberculation in iron pipes in the pH range of 7.5–9.0?

a. Lime
b. Phosphates
c. Sodium bicarbonate
d. Sodium hydroxide

77. Lamella plate or tube settlers with angles of _______ would require backflushing or manual cleaning to remove built-up sludge.

a. 60°
b. 55°
c. 45°
d. 35°

78. The cleaning of microfiltration membranes helps maintain a stable __________ while keeping recovery at or above __________.

a. flux, 95%
b. flux, 99%
c. pressure feed, 95%
d. high pressure feed, 99%

79. The fluoride injection point should be as far away as possible from any _________.

a. phosphate chemicals used for corrosion control
b. polymers used in coagulation
c. calcium, such as lime
d. ferric salts used for coagulation

80. Chlorinating before a microstrainer may cause a sticky, jelly-like coating to form on the microstrainer. What is present in the water that causes this problem?

   a. Soluble iron
   b. Soluble manganese
   c. Slime-producing algae
   d. Filamentous algae or fungi

81. A GAC contactor should be used after __________ when insufficient __________ compounds are not removed.

   a. lime softening, hardness
   b. conventional filtration treatment, synthetic organic
   c. oxidation, natural organic
   d. lime–soda ash softening, hardness

82. Chelation controls __________, and sequestration controls __________.

   a. scale, corrosion
   b. corrosion, scale
   c. scale, scale
   d. corrosion, corrosion

83. What membrane process will remove dissolved solids?

   a. Microfiltration
   b. Reverse osmosis
   c. Nanofiltration
   d. Ultrafiltration

84. Determine the feed rate for alum in mL/min with the following conditions:

| | |
|---|---|
| Plant flow | 16.4 mgd |
| Alum dosage rate | 8.35 mg/L |
| Alum percentage | 48.5% |
| Alum specific gravity | 1.26 |

   a. 139 mL/min of alum
   b. 359 mL/min of alum
   c. 460 mL/min of alum
   d. 589 mL/min of alum

85. What scale prevention chemical can loosen existing deposits and cause red-water complaints?

   a. Silicates
   b. Sodium hexametaphosphate
   c. Carbon dioxide
   d. Sulfuric acid

86. What is the primary purpose of pressure aerators?

   a. To remove volatile gases
   b. To oxidize organic material
   c. To oxidize iron and manganese
   d. To add dissolved oxygen for a more pleasant-tasting water

87. When mechanical equipment is used to dewater alum sludge, it must be pretreated usually using a __________, which is a circular tank equipped with a __________.

a. sludge thickener, stirring mechanism
b. sludge thickener, sludge recirculation feed pump
c. clarifier, chemical feed pump
d. clarifier, stirring mechanism

88. What dictates the amount of coagulant needed?

a. *Giardia* and *Cryptosporidium*
b. Turbidity and organics
c. Turbidity and *Giardia*
d. TOC and *Giardia*

89. What two types of rapid mixing methods have efficiencies directly related to flow?

a. Baffled chambers and static mixers
b. Static mixers and mechanical mixers
c. Static mixers and pumps and conduits
d. Mechanical mixers and baffled chambers

90. In the lime–soda ash softening process, __________ are used to help gather the __________ into particles that will readily settle out of the water.

a. coagulant aids, floc
b. coagulants, fine calcium and magnesium precipitates
c. coagulant aids, precipitates
d. coagulants, floc

91. What will most likely exhaust granular activated carbon first?

a. Taste and odor compounds
b. Organics
c. Chlorine
d. Polymers

92. To make sure the gravel has not shifted or moved up into the sand layer, a metal rod should be used to probe the filter on a grid system to locate and record the depth to gravel. At a minimum, how often should this be done?

a. Every 6 months
b. Once per year
c. Every 2 years
d. Every 3 years

93. What type of caps would be best to use on iron bolts so that the caps become the sacrificial anodes and the iron bolts the protected cathodes?

a. Brass
b. Magnesium
c. Tin
d. Copper

94. A __________ is a conical tank in which the softening reactions take place __________ as the water passes upward through the unit.

    a. clarifier, rather slowly
    b. thickener cone, quite rapidly
    c. softener reactor, rather slowly
    d. pellet reactor, quite rapidly

95. Besides controlling algae by using copper sulfate or powdered activated carbon, what other method has been occasionally used?

    a. Sulfur dioxide
    b. Sodium thiosulfate
    c. Calcium sulfate
    d. Potassium permanganate

96. A 1 ppm alum solution is desired for performing a jar test. If the alum has a specific gravity of 1.25 and is 48.0% alum, how many milliliters are required to make exactly 1,000 mL stock solution?

    a. 1.6 mL of alum
    b. 1.64 mL of alum
    c. 1.65 mL of alum
    d. 1.67 mL of alum

97. At a softening plant, the most severe carryover of sludge in a sedimentation basin occurs when ________.

    a. high winds are blowing in the direction of the weir
    b. the sludge is too high and the flow of water has cut a "river" through it, causing severe short-circuiting
    c. incomplete coagulation–flocculation is occurring
    d. the floc is composed mostly of magnesium hydroxide particles

98. If there is no pretreatment, what membrane processes will remove dissolved organic matter?

    a. Microfiltration and nanofiltration
    b. Nanofiltration and reverse osmosis
    c. Microfiltration and ultrafiltration
    d. Nanofiltration and ultrafiltration

99. According to best practices, tip speed of the last flocculator should be about __________ and operate in the range of __________.

    a. 0.5 ft/s, 0.3–0.8 ft/s
    b. 1.0 ft/s, 0.9–1.3 ft/s
    c. 1.5 ft/s, 1.0–2.0 ft/s
    d. 2.0 ft/s, 1.5–2.5 ft/s

100. Currently, the zebra mussel population is _________.

    a. decreasing in number in most waters in North America
    b. slightly decreasing in number in most waters in the United States
    c. stable due to strong control measures
    d. spreading to most waters of North America

101. In the lime–soda ash process, what will cause increased solubility of taste- and odor-causing compounds?

a. Elevated pH values
b. Elevated temperatures
c. Reactions between carbonate ions and negatively charged organics
d. Ionic bonding of calcium and magnesium carbonates with tannic and fulvic acids

102. Calculate the detention time in hours for the following portion of a treatment plant: Four flocculation basins, each 49.8 ft by 23.9 ft with an average water depth of 10.8 ft; the sedimentation basin is 128 ft long, 79.8 ft wide, and has an average water depth of 10.1 ft; and the flow is 17.2 mgd.

a. 0.87 hr
b. 2.18 hr
c. 4.37 hr
d. 1.61 hr

103. The sediment from presedimentation systems can be kept if samples are collected _________.

a. through core sampling
b. randomly throughout the basin as grab samples
c. as composite sampling throughout the basin
d. and vacuum sealed in glass containers

104. When alum is added to water, it causes the pH to __________, and when ferric salts are added to water, it causes the pH to __________.

a. increase, increase
b. decrease, decrease
c. increase, decrease
d. decrease, increase

105. Lime slakers should be equipped with a __________ that will stop the lime feed before __________.

a. temperature override device, freezing occurs
b. water flow gauge, the lime is over-diluted
c. temperature override device, dangerous temperatures are reached
d. scale, it overfeeds the lime

106. Chlorite monitoring is required only for systems using _________.

a. ozone
b. sodium hypochlorite
c. chloramines
d. chlorine dioxide

107. A 5-min drawdown test result showed that 635 mL of alum was being injected to treat the raw water. The alum was 10.38 lb/gal. If the plant is treating 3,440 gpm, what is the alum dosage in mg/L?

a. 9.13 mg/L of alum
b. 13.8 mg/L of alum
c. 18.8 mg/L of alum
d. 19.7 mg/L of alum

108. The ideal alkalinity for manganese greensand filters for the removal of iron and manganese is __________, as $CaCO_3$.

    a. >75 mg/L
    b. >100 mg/L
    c. >110 mg/L
    d. >120 mg/L

109. What is an objective of a corrosion control program?

    a. To reduce lead and copper
    b. To provide stable water
    c. To reduce corrosion and scaling
    d. All the above

110. When ozone reacts with organic contaminants to form organic acids, what may this eventually contribute to the distribution system?

    a. Biofouling
    b. High carbon gases
    c. Nitrification
    d. Tubercles

111. Under favorable conditions, combined chlorine, such as monochloramine and dichloroamine, compared to the bactericidal action of free chlorine would require __________ times more combined chlorine residual.

    a. 4
    b. 11
    c. 14
    d. 25

112. Slow sand filters are not normally used with source waters that consistently have turbidities above _________.

    a. 10 ntu
    b. 15 ntu
    c. 20 ntu
    d. 25 ntu

113. What percent hypochlorite solution would result if 275 gal of a 5.8% solution were mixed with 195 gal of a 15% solution? Assume both solutions have the same density.

    a. 3.4% final solution
    b. 6.2% final solution
    c. 9.6% final solution
    d. 11.2% final solution

114. If a reverse osmosis system is to be shut down for a period of time, membranes should be flushed with __________ or __________.

    a. acidified feedwater, unchlorinated product water
    b. finished water, a cleaning solution
    c. acidified feedwater, a cleaning solution
    d. unchlorinated water, a cleaning solution

115. The precipitative softening process uses _________ that render __________ calcium and magnesium __________.

a. chemical reactions, dissolved, insoluble
b. chemical reactions, colloidal, to precipitate
c. physical processes, dissolved, insoluble
d. physical processes, colloidal, to precipitate

116. How much sludge will 1 mg/L of dry-basis alum produce if it consumes 0.5 mg/L alkalinity as $CaCO_3$?

a. 0.15 mg/L
b. 0.26 mg/L
c. 0.50 mg/L
d. 0.62 mg/L

117. Chelation is a chemical treatment process used to control _________.

a. scale formation
b. corrosion by products
c. iron and manganese precipitation
d. tubercle precipitates

118. What is the typical detention time for the recarbonation process?

a. 5–10 min
b. 10–15 min
c. 15–30 min
d. 30–60 min

119. A limestone contactor is a treatment device that _________.

a. uses powdered limestone
b. increases pH and alkalinity
c. is complex and costs much more than lime application equipment
d. requires frequent maintenance

120. Some iron species under certain oxidation conditions are so fine that they penetrate the filter. What can be done to prevent this?

a. Increase oxidation detention time
b. Add a filter aid
c. Add a flocculant after oxidation
d. All of the above

121. Ammonia cylinders are more commonly used than one-tons as with chlorine because ammonia is fed at about __________ of the amount of chlorine used.

a. 20% to 25%
b. 25% to 33%
c. 30% to 40%
d. 40% to 50%

122. Find the amount of iron and manganese removed per year from a plant that treats an average of 10.88 mgd if the average iron concentration is 0.91 ppm and the average manganese concentration is 0.15 ppm. The iron removal efficiency is 89.5%, and for the manganese, it is 71.7%.

 a. 26,100 lb/yr of Fe removed and 3,450 lb/year of Mn removed
 b. 26,500 lb/yr of Fe removed and 3,500 lb/year of Mn removed
 c. 27,000 lb/yr of Fe removed and 3,600 lb/year of Mn removed
 d. 28,000 lb/yr of Fe removed and 3,750 lb/year of Mn removed

123. What makes ozone more stable in the water?

 a. Low alkalinity
 b. High alkalinity
 c. High temperatures
 d. Low oxygen levels

124. A gravity thickener would include _________.

 a. a belt press
 b. freeze-thawing beds
 c. cartridges
 d. vacuum filtration

125. What is the best solution for iron problems when using exchange resins?

 a. Add a strong reducing agent
 b. Add a strong reducing agent with polyphosphates
 c. Add polyphosphates
 d. Remove iron before the exchange resin process

126. The most common higher-rate filter media consists of the following (from bottom to top):

 a. Sand, silica sand, and anthracite
 b. Garnet sand, silica sand, and anthracite
 c. Silica sand, garnet sand, gravel, and anthracite
 d. Cobbles, gravel, silica sand, and anthracite

127. Corrosion will most probably occur when the __________ is (are) low.

 a. dissolved oxygen
 b. alkalinity
 c. total dissolved solids
 d. temperature

128. Concentration cell corrosion is a form of localized corrosion that can form _________.

 a. deep pitting and tubercles at the anode
 b. deep pitting at the anode and tubercles at the cathode
 c. tubercles at the anode and deep pitting at the cathode
 d. deep pitting and tubercles at the cathode

129. What will sequester iron but will not prevent corrosion because it only masks the corrosion by products?

a. EDTA
b. Silicate
c. Zinc orthophosphate
d. Polyphosphates

130. What type of aerators are equipped with a submersible pump?

a. Mechanical aerators
b. Draft-tube aerators
c. Diffuser aerators
d. Packed towers

131. What chemical used for oxidizing iron and manganese is the second most powerful oxidant in its powerful biocidal efficacy and will not produce organic by products?

a. Chlorine dioxide
b. Potassium permanganate
c. Chlorine
d. Ozone

132. What softening process has the advantages of producing less sludge and being easier to store, feed, and handle than lime?

a. Split treatment
b. Excess-soda ash softening
c. Composite softening
d. Caustic-soda treatment

133. What does a streaming current detector measure on a continuous basis?

a. The negative and positive charges
b. The net positive charge
c. The colloidal surface charge
d. The net ionic charge

134. If fluorosilicic acid must be diluted, never dilute it with water to acid in the range of _________.

a. 4:1 to 8:1.
b. 5:1 to 10:1.
c. 8:1 to 12:1.
d. 10:1 to 20:1.

135. Typically, air scour will last __________ min, followed by water alone at a rate of __________.

a. 2–3, 3–5 gpm/ft$^2$
b. 3–7, 5–7 gpm/ft$^2$
c. 2–3, 5–7 gpm/ft$^2$
d. 3–7, 3–5 gpm/ft$^2$

136. Calculate the CT and inactivation ratio for a water treatment plant given the following parameters and determine whether this treatment facility meets the CT.

Daily parameters:

Detention time: 130 min

pH: 7.4

Temperature: 18°C

Lowest chlorine residual: 0.50 mg/L

A 2.0-log removal is required for this system.

a. 1.84 is the inactivation ratio; no
b. 0.53 is the inactivation ratio; no
c. 1.86 is the inactivation ratio; yes
d. 1.87 is the inactivation ratio; yes

137. A 10-min drawdown test result showed that 343 mL of a cationic polymer was being used to treat raw water. The specific gravity of the polymer is 1.27. If the plant is treating 8,470 gpm, what is the polymer dosage in mg/L?

a. 1.35 mg/L of polymer
b. 1.36 mg/L of polymer
c. 1.37 mg/L of polymer
d. 1.40 mg/L of polymer

138. What increases faster in a conventional sand filter than in a higher-rate filter?

a. Penetration of floc
b. Organics causing color
c. Head loss
d. Metal precipitates

139. Find the total bicarbonate, carbonate, and hydroxide alkalinity for a water sample with the following characteristics:

| | | Titration Result |
|---|---|---|
| Phenolphthalein (P) alkalinity | = | 32 mg/L |
| Total (T) alkalinity | = | 58 mg/L |

Use the following table to solve this problem:

| Alkalinity, mg/L as $CaCO_3$ | | | |
|---|---|---|---|
| Results of Titration | Bicarbonate Alkalinity | Carbonate Alkalinity | Hydroxide Alkalinity |
| P = 0 | T | 0 | 0 |
| P is less than ½T | T–2P | 2P | 0 |
| P = ½T | 0 | 2P | 0 |
| P is greater than ½T | 0 | 2T–2P | 2P–T |
| P = T | 0 | 0 | 0 |

Note: P = phenolphthalein alkalinity and T = total alkalinity

a. 0 bicarbonate alkalinity, 50 mg/L as carbonate alkalinity, 6 mg/L as hydroxide alkalinity
b. 0 bicarbonate alkalinity, 52 mg/L as carbonate alkalinity, 7 mg/L as hydroxide alkalinity
c. 0 bicarbonate alkalinity, 52 mg/L as carbonate alkalinity, 6 mg/L as hydroxide alkalinity
d. 0 bicarbonate alkalinity, 52 mg/L as carbonate alkalinity, 7 mg/L as hydroxide alkalinity

140. Plate and tube settlers occupy an area that is as little as__________ of a typical open basin.

a. 5%
b. 10%
c. 20%
d. 25%

141. When performing a jar test to determine PAC dosage, how much sample should be discarded before performing the threshold odor test?

a. 100 mL
b. 200 mL
c. 300 mL
d. 400 mL

142. The problems with granular activated carbon (GAC), such as chlorine or synthetic organic chemicals exhausting it prematurely, fouling, and other issues, can be avoided by ________.

a. installing a GAC contactor
b. optimizing the coagulation/flocculation process
c. stopping prechlorination
d. installing an aerator to reduce tastes and odors

143. Because conventional sedimentation basins can have short-circuiting or be poorly designed, what is a common way to determine the operation of the sedimentation basin?

a. Sludge settling
b. Weir overflow rate
c. Floc carryover
d. Depth of sludge from top of effluent weir

144. When are granular activated carbon contactors backwashed?

a. When organic acids breakthrough begins at the effluent
b. When silica breakthrough begins at the effluent
c. When there is a noticeable pressure drop between inlet and outlet
d. After a specific number of gallons have been treated

145. The type of pipe feeding a powdered activated carbon slurry should be __________ material, and the day tanks are typically made of __________.

a. metal, fiberglass
b. rubber or plastic, plastic
c. corrosion- and erosion-resistant, fiberglass
d. plastic and corrosion-resistant, plastic

146. *Giardia* and *Cryptosporidium* are typically removed by __________. __________ can be inactivated by sufficient $C \times T$ values employing chlorine, but __________ is considered to be resistant to chlorine.

  a. coagulation, flocculation, and sedimentation; *Giardia; Cryptosporidium*
  b. coagulation, flocculation, and sedimentation; *Cryptosporidium; Giardia*
  c. filtration, *Giardia, Cryptosporidium*
  d. filtration, *Cryptosporidium, Giardia*

147. Why can't lime sludge be discharged to a sanitary sewer system?

  a. High pH will kill the activated sludge at the sewage plant.
  b. It will corrode the pumps at the sewer plant as it is highly caustic.
  c. Treatment of this type of sludge will incur high additional cost.
  d. Deposits will build up and block the pipes to the sewage plant.

148. The principal use of granular activated carbon contactors when treating groundwater is to remove _________.

  a. inorganic contaminants
  b. organic contaminants
  c. metal contaminants
  d. synthetic organic chemicals

149. What process would be best for removing oils produced by algae or industrial chemicals?

  a. Granular activated carbon
  b. Packed tower aeration
  c. Cone aerator
  d. Slat-and-coke tray aerator

150. The harmful effects of rate increase can be reduced with ________.

  a. filter aids
  b. enhanced coagulation
  c. adjusting pH to the optimum pH of the coagulant being used
  d. lime softening

151. What is the psi at the bottom of a polymer storage tank if the level of the polymer is 9.66 ft? The density of the polymer is 10.34 lb/gal.

  a. 5.18 psi
  b. 77.3 psi
  c. 100 psi
  d. 747 psi

152. What percent hypochlorite solution would result if 275 gal of a 12.5% solution were mixed with 110 gal of a 5.2% solution? Assume both solutions have the same density and are measured to the nearest gallon (thus, they both have three significant figures).

  a. 7.3% final solution
  b. 8.9% final solution
  c. 10.4% final solution
  d. 10.8% final solution

153. When both iron and manganese are treated using manganese greensand filters, the chlorine feed for iron should be __________ times the iron concentration and __________ times the manganese concentration.

a. 1.0, 1.5
b. 1.0, 2.0
c. 1.5, 2.0
d. 1.5, 2.5

154. To prevent $CO_2$ from being corrosive, its concentration in finished water should be below _________.

a. 5 mg/L
b. 7 mg/L
c. 9 mg/L
d. 11 mg/L

155. Recarbonation of water that was __________ in magnesium will usually use a __________ process.

a. low, two-stage
b. high, three-stage
c. low, three-stage
d. high, two-stage

156. Post treatment for a reverse osmosis unit may be required to remove _________.

a. radon
b. $H_2S$
c. sodium ions
d. zinc

157. What chemical contactor devices used in corrosion control are simple, low-cost, and require little maintenance, making them especially suitable in small water systems?

a. Sodium silicate contactors
b. Phosphate contactors
c. Limestone contactors
d. Sodium bicarbonate contactors

158. What will occur if fluorosilicic acid is diluted in the range of 10:1 to 20:1?

a. Sodium fluoride crystals may form.
b. A fluoride gel may form.
c. An insoluble silica precipitate often forms.
d. Sodium silicate gel often forms.

159. What are the key factor(s) in the operation of a solid-contact basin?

a. Temperature and flow
b. Temperature and chemical coagulant
c. Mixing speed and sludge blanket
d. Flow

160. What method of flow control is accomplished mechanically?

    a. Proportional-level equal rate control
    b. Variable-level influent flow splitting
    c. Proportional-level flow splitting
    d. Head pressure level flow control

161. What are the basic water quality control tests used in operating the aeration process?

    a. pH and temperature
    b. Dissolved oxygen and pH
    c. Dissolved oxygen, carbon dioxide, and pH
    d. pH, temperature, and dissolved oxygen

162. Reverse osmosis can remove the following:

    a. Metal ions
    b. Aqueous salts
    c. Sugar
    d. All of the above

163. A common mistake made in the flocculation basins is _________.

    a. adding the coagulant aid in the first flocculator unit
    b. to keep too many units in service
    c. to taper the mixing gradient too steeply
    d. to taper the mixing gradient too gently

164. The most common water flow on softening exchangers are __________ because they can operate __________.

    a. upward, longer when the beads are in a slight suspension
    b. downward, longer when the beads are tightly packed
    c. upward, at a higher flow rate
    d. downward, at a higher flow rate

165. What chemical used for oxidizing iron and manganese produces organic by products, such as carboxylic acids and aldehydes?

    a. Chlorine
    b. Chlorine dioxide
    c. Ozone
    d. Potassium permanganate

166. Diatomaceous earth filtration is not effective against _________.

    a. *Giardia*
    b. *Cryptosporidium*
    c. color
    d. algae

167. In the reverse osmosis (RO) process, pretreatment of feedwater is usually necessary to prevent premature __________ of the RO membrane.

    a. backwashing
    b. deterioration
    c. fouling
    d. need for a flux process

168. If a water plant is feeding a sodium fluoride solution, the water used to make the solution should be softened if it exceeds __________ as calcium carbonate.

a. 50 mg/L
b. 75 mg/L
c. 100 mg/L
d. 110 mg/L

169. If lagoons are used to dewater sludge with the water being decanted, the use of the lagoon should be discontinued when the depth of the sludge reaches _________.

a. 1–3 ft
b. 3–5 ft
c. 5–7 ft
d. 7–9 ft

170. The bed depth of a filter is measured from the top of the media to the __________ with a __________ steel rod.

a. bottom of the gravel bed, $^3/_8$-in.
b. top of the gravel bed, $^3/_8$-in.
c. bottom of the gravel bed, $^3/_4$-in.
d. top of the gravel bed, $^3/_4$-in.

171. When excess lime is used to remove __________ hardness, then __________ is required to __________ of the water.

a. carbonate, aeration, lower the pH
b. carbonate, aeration, raise the pH
c. noncarbonate, recarbonation, lower the pH
d. noncarbonate, recarbonation, raise the pH

172. A 25% caustic solution with a specific gravity of 1.278 is being used to raise the pH. If the flow from the clearwell is currently set at 12.8 ft$^3$/s and the caustic feed is 52 mL/min, what is the dosage of caustic in mg/L?

a. 0.71 mg/L of caustic
b. 0.72 mg/L of caustic
c. 0.74 mg/L of caustic
d. 0.76 mg/L of caustic

173. What chemical used for oxidizing iron and manganese has a high cost in equipment, operation, and maintenance, and if excessive quantities are used, it can oxidize manganese to permanganate and cause pink water?

a. Chlorine dioxide
b. Calcium hypochlorite
c. Ozone
d. Potassium permanganate

174. Pleated membrane technology is designed to serve as a __________ barrier for high-quality waters with low fouling potential.

a. virus
b. bacteria
c. *Giardia*
d. *Cryptosporidium*

175. What will most likely reduce the capacity of powdered activated carbon particles to absorb?

   a. Cationic monovalent metals
   b. Cationic multivalent metals
   c. Nitrates and nitrites
   d. Coagulants and chlorine

176. Typically, negative head occurs in filters that have a water depth over the unexpanded filter bed that is less than _________.

   a. 3.5 ft
   b. 4.0 ft
   c. 4.5 ft
   d. 5.0 ft

177. When the total dissolved solids concentration __________, the solubility of calcium carbonate __________.

   a. decreases, increases
   b. decreases, decreases
   c. increases, increases
   d. increases, decreases

178. Adding chlorine before granular activated carbon (GAC) to kill bacteria is __________. The concentration of bacteria in a GAC filter is __________ than in the influent.

   a. very effective, many times smaller
   b. somewhat effective, only slightly better
   c. ineffective, hundreds of times higher
   d. ineffective, thousands of times higher

179. A naturally occurring chelating agent is _________.

   a. zeolites
   b. humic acid
   c. sodium glyoxylate
   d. glyceraldehyde

180. When using calcium hypochlorite, calcium carbonate scaling can occur in the suction and discharge side of the pump as well as in the pump. What would be the best solution to this problem?

   a. Use a two-tank setup to eliminate the lime sludge being sucked in from the bottom of the first tank
   b. Clean the lines and pump with muriatic acid
   c. Install new lines and clean the pump with dilute HCl
   d. Use deionized or soft water as carrier water

181. Ozone may convert large organic molecules into smaller ones. These are called biodegradable dissolved organic carbon (BDOC) compounds. These BDOC compounds serve as "food" for _________.

   a. bacteria
   b. saprophytes
   c. ciliated protozoans
   d. flagellated protozoans

182. How long will it take in hours and minutes for a pump to lower the water level in a tank from its current value to 10.0 ft to another tank at a higher level given the following data? Disregard the change in water level while pumping.

Hp = 60.5
Total head = 385 ft
Pump efficiency = 81.3%
Live 4–20 mA signal = 16.07
Tank radius = 35.5 ft
Tank capacity = 38.5 ft
Influent flow = 0

a. 18 hr and 34 min
b. 18 hr and 40 min
c. 18 hr and 44 min
d. 18 hr and 58 min

183. The oxidation of manganese in surface water is not effective until the pH goes to or surpasses _________.

a. 9
b. 9.2
c. 9.5
d. 9.8

184. In coagulation, once the zeta potential forces are reduced below the __________, the particles in suspension will start to coalesce.

a. molecular forces
b. van der Waals forces
c. ionic forces
d. covalent forces

185. How often are the maximum daily temperatures determined to establish the optimal fluoride levels for any particular state?

a. Monthly
b. Annually
c. A running annual average
d. Every 5 years

186. Water becomes capable of ___________ iron and manganese when the oxygen levels __________.

a. precipitating, increase
b. dissolving, increase
c. precipitating, decrease
d. dissolving, decrease

187. When using chlorine disinfectants, lower temperatures favor the formation of __________. Lower temperatures also show __________ to persist somewhat longer.

a. HOCl, chlorine residuals
b. HOCl, –OCl
c. –OCl, HOCl
d. –OCl, –OCl

188. What will a coagulant aid not achieve from the following?

    a. Build stronger, more settleable floc
    b. Overcome temperature increases and drops
    c. Reduce the amount of coagulant needed
    d. Reduce the amount of sludge produced

189. As the Total Organic Carbon (TOC) increases, the TOC percentage removal required from a source water supply ________.

    a. increases as alkalinity increases
    b. increases as alkalinity decreases
    c. decreases as pH increases
    d. decreases as temperature increases

190. What membrane process would be used to treat brackish waters?

    a. Nanofiltration
    b. Reverse osmosis
    c. Ultrafiltration
    d. Microfiltration

191. A Leopold filter bottom is a patented filter underdrain system using a series of perforated __________ with channels to carry the water.

    a. PVDF pipes
    b. CPVC pipes
    c. vitrified clay blocks
    d. ceramic diffusers

192. In a granular activated carbon filter, __________ can drive previously deposited material through the filter and can form __________ through the filter bed.

    a. a broken underdrain nozzle, channels
    b. excess backwashing, cracks
    c. flow fluctuations, channels
    d. large mudballs, cracking

193. What type of pretreatment settling units can use coagulants because it has been shown to be very effective?

    a. Lamella plates
    b. Tube settlers
    c. Plate settlers
    d. Cyclone degritters

194. Continuous regeneration (CR) is usually used in greensand filters when __________ is mostly being removed and intermittent regeneration (IR) when __________ is mainly being removed. The __________ regeneration is used while the filter is offline.

    a. iron, manganese, IR
    b. manganese, iron, CR
    c. iron, manganese, CR
    d. manganese, iron, IR

195. Because the gas generated from __________ is pure $CO_2$, the pipes for the diffusers can be __________ than those needed for $CO_2$ generated on site by combustion.

a. liquid or gas $CO_2$ cylinders, smaller
b. liquid or gas $CO_2$ cylinders, larger
c. dry ice or liquid $CO_2$, smaller
d. dry ice or liquid $CO_2$, larger

196. As the pH of a particular water sample increases, the calcium carbonate level will __________, and as the temperature __________, the solubility of the calcium carbonate __________.

a. increase, increases, decreases
b. decrease, increases, decreases
c. increase, decreases, increases
d. decrease, decreases, increases

197. Beneath the slime layer of certain iron bacteria in a pipe, what is produced that reduces the pH and speeds up the corrosion rate?

a. $H_2S$
b. $H_2$
c. $CO_2$
d. $SO_4$

198. The process of aeration used in packed towers is __________.

a. adsorption
b. absorption
c. desorption
d. degasification

199. What does $C \times T$ stand for?

a. Contact times time
b. Chlorine concentration contact time
c. Chlorine times temperature
d. Chlorine concentration multiplied by the time of contact

200. A procedure calls for diluting a 12.5% sodium hypochlorite (NaOCl) solution to a 5.25% solution. If exactly 1,000 gal of the 5.25% solution is desired, how much of the 12.5% solution is required?

a. 420 gal of the 12.5% NaOCl solution are needed
b. 425 gal of the 12.5% NaOCl solution are needed
c. 520 gal of the 12.5% NaOCl solution are needed
d. 580 gal of the 12.5% NaOCl solution are needed

201. A 5-min drawdown test result showed that 232 mL of a cationic polymer was being injected. The specific gravity of the polymer is 1.28. If the plant is treating 6,080 gpm, what is the polymer dosage in mg/L?

a. 2.57 mg/L of polymer
b. 2.58 mg/L of polymer
c. 2.59 mg/L of polymer
d. 2.6 mg/L of polymer

202. The ripening of a filter after it backwashes depends on __________.

    a. the attraction of sticky particles to the anthracite or sand and then those attached particles attracting even more sticky particles until the filter is ripened
    b. the bed slowly settling and compacting, which prevents turbidity breakthrough
    c. the sand collecting particles first by straining action; then once anthracite starts collecting particles, the filter is ripened
    d. fluid drag jostling the irregular-shaped anthracite particles into positions that eventually prevent most turbidity breakthrough

203. Where can powdered activated carbon be applied in a conventional water treatment plant?

    a. Before coagulation
    b. Before the filters
    c. Before sedimentation
    d. All the above

204. Predict the amount of sludge a water plant will produce per day if it is treating 6.7 mgd with 14 mg/L of ferric chloride and the plant is adding 5 mg/L carbon for taste and odor control. The source water turbidity is averaging 11 ntu, and the historic B value for this plant is 0.9.

    a. 3,000 lb/d of sludge
    b. 3,050 lb/d of sludge
    c. 3,100 lb/d of sludge
    d. 3,110 lb/d of sludge

205. The best way to remove $H_2S$ is by scrubbing at about __________ by bubbling $CO_2$ through the water before aeration to reduce the pH.

    a. pH 4.5
    b. pH 5.5
    c. pH 6.0
    d. pH 6.5

206. The optimum pH for the removal of excess fluoride by activated alumina is __________.

    a. 5.5
    b. 6.5
    c. 7.0
    d. 7.5

207. Producers of exchange resins are capable of engineering many types of resins, including resins capable of controlling __________ and __________.

    a. functioning chemical matrices, cross-linking
    b. functional chemical groups, exchange sites
    c. exchange sites, cross-linking
    d. exchange capacity, porosity

208. A polymer solution weighs 1.26 g/mL. The water treatment plant is using 89.3 mL/min of the polymer for treating 30.1 $ft^3$/s. What is the polymer dosage?

a. 0.45 mg/L of polymer
b. 2.20 mg/L of polymer
c. 11.85 mg/L of polymer
d. 18.3 mg/L of polymer

209. At the optimum pH level for the removal of fluoride by activated alumina, the fluoride can be reduced to below __________.

a. 0.1 mg/L
b. 0.3 mg/L
c. 0.5 mg/L
d. 1.0 mg/L

210. Although __________ can be added to lime–soda ash rapid-mix basins, better results are usually obtained if they are added __________ in a separate rapid-mix basin.

a. coagulants, upstream
b. coagulants, downstream
c. disinfectants, upstream
d. disinfectants, downstream

211. What is (are) the potential concerns when ozone is used as a primary disinfectant?

a. Chlorate and chlorite production
b. Bromate and formaldehyde production
c. Haloacetic acids formation
d. Bromate and chlorite formation

212. An organic compound from components of pipe, coatings, and joint adhesives would most likely include __________.

a. polychlorinated biphenyls
b. acrylamide
c. polynuclear aromatic hydrocarbons
d. polyvinyl chloride

213. Determine the hydrated lime, soda ash, and carbon dioxide dose requirements in milligrams per liter for water with the following characteristics:

| | Source Water | Softened Water |
|---|---|---|
| Total alkalinity, mg/L | 286 mg/L as $CaCO_3$ | 76 mg/L |
| Total hardness, mg/L | 412 mg/L as $CaCO_3$ | 81 mg/L |
| $CO_2$, mg/L | 25 mg/L | 0 mg/L |
| $Mg^{2+}$ | 33 mg/L | 6.5 mg/L |
| pH | 7.5 | 8.8 |
| Lime purity | 92.5% | |

Use the following table to solve this problem:

Calculations for chemical precipitation softening process

| Molecular weights of chemical compounds | |
|---|---|
| Compound | Molecular Weight |
| Alkalinity, as $CaCO_3$ | 100.1 |
| Carbon dioxide, $CO_2$ | 44.0 |
| Hardness, as $CaCO_3$ | 100.1 |
| Hydrated lime, $Ca(OH)_2$ | 74.1 |
| Magnesium, $Mg^{2+}$ | 24.3 |
| Magnesium hydroxide, $Mg(OH)_2$ | 58.3 |
| Quicklime, CaO | 56.1 |
| Soda ash, $Na_2CO_3$ | 106.0 |

a. 360 mg/L of $Ca(OH)_2$; 47.7 mg/L of soda ash
b. 360 mg/L of $Ca(OH)_2$; 48 mg/L of soda ash
c. 414 mg/L of $Ca(OH)_2$; 47.7 mg/L of soda ash
d. 414 mg/L of $Ca(OH)_2$; 48 mg/L of soda ash

214. What is the typical backwash rate for resin beds?

a. 3–5 gpm/ft$^2$
b. 5–8 gpm/ft$^2$
c. 7–10 gpm/ft$^2$
d. 11–14 gpm/ft$^2$

215. A plant is treating water at a rate of 10.8 ft$^3$/s. If lime is being added at a rate of 100.2 g/min, what are the lime usage in lb/d and the dosage in mg/L?

a. 5.4 mg/L of lime
b. 5.43 mg/L of lime
c. 5.45 mg/L of lime
d. 5.46 mg/L of lime

216. What chemical is most likely to cause taste and odor problems in finished water?

a. Free chlorine
b. Hypochlorous acid
c. Hypochlorite ion
d. Dichloroamine

217. A treatment plant is adding 162.8 g/min of soda ash to its treated water. If the plant is producing water at 7.44 mgd, what is the soda ash usage in lb/d and dosage in mg/L?

a. 8.27 mg/L of soda ash
b. 8.32 mg/L of soda ash
c. 8.40 mg/L of soda ash
d. 8.44 mg/L of soda ash

## Laboratory Analysis

1. There are two types of acidity: phenolphthalein acidity, which is prevalent __________, and mineral acid acidity, which tends to be in the _________.

   a. below pH 4.3, lower pH ranges
   b. above pH 4.3, lower pH ranges
   c. below pH 8.3, range of pH 4.3–8.3
   d. above pH 8.3, range of pH 4.3–8.3

2. What is carbonate alkalinity?

   a. Alkalinity caused by $CO_3^{-2}$
   b. Alkalinity caused by $HCO^{-3}$
   c. Alkalinity caused by $CaCO_3$
   d. Alkalinity caused by $CaCO_3$ and $H_2CO_3$

3. The platinum–cobalt method is a procedure used to determine the amount of _________ in water.

   a. arsenic
   b. total organic halogens
   c. color
   d. iron

4. Chlorine dioxide is _________ times more powerful than chlorine.

   a. 1.2–1.8
   b. 2.0–2.5
   c. 2.7–3.1
   d. 3.2–3.7

5. What treatment chemical will raise the pH?

   a. Chlorine
   b. Sodium hypochlorite
   c. Alum
   d. Ferric sulfate

6. A calcium sulfate molecule would be considered a ________.

   a. colloidal solid
   b. suspended solid
   c. dissolved solid
   d. molecular solid

7. Regarding the USEPA sampling frequency for radionuclides in drinking water for individual contaminants at each entry point: If results of testing are equal to or greater than one-half the MCL and equal to or less than the MCL, what is the frequency of sampling?

   a. 3 years
   b. 6 years
   c. 9 years
   d. 12 years

8. Mineral acids are measured by titration to a pH of __________ using __________ as an indicator.

   a. 4.3, methyl red
   b. 4.3, methyl orange
   c. 8.3, methyl red
   d. 8.3, methyl orange

9. In groundwater systems that disinfect and have any uncorrected significant deficiencies or draw water from a susceptible source, 4-log inactivation of __________ must be demonstrated by providing __________.

   a. *Cryptosporidium,* that coliforms are absent in 95% or more of all samples collected for that sampling period
   b. *Cryptosporidium,* adequate $C \times T$
   c. viruses, that coliforms are absent in 95% or more of all samples collected for that sampling period
   d. viruses, adequate $C \times T$

10. What are the three forms of carbonate hardness?

   a. $CaCO_3$, $HCO_3^{-1}$, and $CO_3^{-2}$
   b. OH, $HCO_3^{-1}$, and $CO_3^{-2}$
   c. $CaCO_3$, $MgCO_3$, and $CaSO_4$
   d. None of the above

11. When measuring chlorine residual using the amperometric method, be aware that the electrodes are susceptible to interferences from ________.

   a. calcium
   b. nitrite
   c. dissolved oxygen
   d. organics

12. A __________ aliquot of sample for conductivity measurement may be collected from __________ sample collected from a regularly used bacteriological sample point or from a sample collected __________.

   a. 50 mL, an unpreserved, at the entry point to the distribution system
   b. 100 mL, a preserved, at the entry point to the distribution system
   c. 50 mL, an unpreserved, from the furthest points in the distribution system
   d. 100 mL, an unpreserved, from the furthest points in the distribution system

13. A specific-ion meter resembles a __________ except for the addition of a __________ and a provision for a selective-ion electrode.

   a. pH meter, milliamp scale
   b. pH meter, millivolt scale
   c. electrophotometer, milliamp scale
   d. electrophotometer, millivolt scale

14. Probably the best and easiest method to transport samples to a laboratory is by mail except for microbiological and ________.

   a. volatile organic compounds
   b. certain radiological samples
   c. certain synthetic organic compounds
   d. certain pesticides

15. What test method for coliform bacteria has three distinct steps: the presumptive, confirmed, and completed test?

    a. MF
    b. MTF
    c. MNO-MUG
    d. Both a and b

16. Disinfection byproducts increase as _________ increases, and low pH favors __________, and high pH favors __________.

    a. pH, HAA5, TTHM
    b. pH, TTHM, HAA5
    c. temperature, TTHM, HAA5
    d. temperature, HAA5, TTHM

17. What information can particle counters provide?

    a. Size and number
    b. Qualitative measurement
    c. Number of bacteria
    d. The number of particles smaller than can be counted by turbidity meters

18. When samples are collected for analyses of haloacetic acids 5 (HAA5), what is added to the sample container to quench more HAA5 formation?

    a. $HNO_3$
    b. $NH_4Cl$
    c. Dilute $H_2SO_4$
    d. $NO_2$

19. The reason that a sample has a holding time is usually because the sample's _________.

    a. contaminant changes for various reasons
    b. preservative degrades over time
    c. time to report the result passes the deadline
    d. preservative reacts with the contaminate in question or other constituents in the sample over time

20. If a routine or repeat sample tests positive for total coliforms, it must also be analyzed for _________.

    a. repeat coliform test and fecal coliforms
    b. fecal coliforms
    c. fecal coliforms and *E. Coli*
    d. *E. Coli*

21. A gas chromatograph–mass spectrometer would be used to determine __________ in a sample.

    a. metals
    b. inorganic compounds
    c. organic compounds
    d. elements

22. Regarding the USEPA sampling frequency for radionuclides in drinking water for individual contaminants at each entry point: If results of testing are below the detection limit of an approval method, what is the frequency of sampling?

    a. 3 years
    b. 6 years
    c. 9 years
    d. 12 years

23. In the confirmed test using brilliant green lactose broth, if __________ been produced after __________ of incubation, the test is negative.

    a. yellow colonies have, 24 hours
    b. yellow colonies have, 48 hours
    c. no gas has, 24 hours
    d. no gas has, 48 hours

24. What is bicarbonate alkalinity?

    a. Alkalinity caused by $CO_3^{-2}$
    b. Alkalinity caused by $HCO^{-3}$
    c. Alkalinity caused by $CaCO_3$
    d. Alkalinity caused by $CaCO_3$ and $H_2CO_3$

25. Adjustments to the carbon dioxide dosage in recarbonation should be based on both the __________ and results of __________.

    a. total hardness tests, jar tests
    b. alkalinity tests, coupon testing
    c. Langelier saturation index, coupon testing
    d. Langelier saturation index, jar tests

26. Another method for the modified Winkler method is the __________.

    a. EDTA titration method
    b. modified EDTA titration method
    c. modified Mohr method
    d. iodometric method

27. In the presumptive test, if gas is produced after either the __________ or __________ incubation period, the sample must undergo the __________ test.

    a. 24-hr, 36-hr, completed
    b. 24-hr, 48-hr, confirmed
    c. 24-hr, 36-hr, final
    d. 48-hr, 72-hr, completed

28. The solubility of calcium carbonate in water __________ as temperature increases, and the TDS concentration __________ as the solubility of calcium carbonate increases.

    a. decreases, increases
    b. increases, increases
    c. decreases, decreases
    d. increases, decreases

29. Most particles in water are negatively charged, and thus they repel each other. What is this repelling electrical force called?

    a. van der Waals force
    b. Zeta potential
    c. Weak electromagnetic force
    d. Ionic potential

30. What is the correct chemical formula for hydrated ferric sulfate?

    a. $Fe(SO_4)_2 \cdot 2H_2O$
    b. $Fe(SO_4)_3 \cdot 3H_2O$
    c. $Fe_2(SO_4)_3 \cdot 3H_2O$
    d. $Fe_2(SO_4)_4 \cdot 4H_2O$

31. All __________ and gas producing samples from the presence–absence presumptive stage test must then be confirmed as described for the multiple-tube fermentation-confirmed step using __________ tubes.

    a. pink, lauryl tryptose broth (LTB)
    b. yellow, brilliant green bile (BGB)
    c. pink, BGB
    d. yellow, LTB

32. When dissolved oxygen increases, the corrosion rate __________, and when the pH and alkalinity increase, the corrosion rate __________.

    a. increases, increases
    b. decreases, decreases
    c. decreases, increases
    d. increases, decreases

33. In the presumptive test, if coliforms are present in the water, the __________ they produce will begin to form __________ in each inverted vial within a __________ period.

    a. gas, bubbles, 24-hr
    b. gas, bubbles, 48-hr
    c. colonies, a metallic sheen, 24-hr
    d. colonies, a metallic sheen, 48-hr

34. What is a good laboratory material in which to store dry chemical reagents but not for collecting samples?

    a. Kimax
    b. Pyrex
    c. Soft glass
    d. Plastic

35. Desiccators are used to __________ material or items before __________, and they also provide a __________ environment.

    a. rapidly cool, weighing, moisture-free
    b. rapidly cool, analyses, dust- and moisture-free
    c. slowly cool, weighing, dust- and moisture-free
    d. slowly cool, analyses, moisture-free

36. Water that has __________ will be very corrosive if it becomes __________.

    a. high alkalinity, anaerobic
    b. low alkalinity, saturated with phosphates
    c. high alkalinity, aerobic
    d. low alkalinity, saturated with oxygen

37. If any stains or chemical residues remain after normal cleaning of labware, the glassware should be washed with a cleaning solution, such as ________.

    a. hydrogen peroxide
    b. 10% hydrochloric acid solution
    c. 2*N* solution of hydrogen sulfate
    d. acid-dichromate

38. __________ have a __________ light source and provide a __________ measurement but cannot detect particles that do not scatter light.

    a. Particle counters, fixed, qualitative
    b. Particle counters, movable, quantitative
    c. Turbidity meters, fixed, qualitative
    d. Turbidity meters, movable, quantitative

39. In the multiple-tube fermentation method regarding the completed test, a small portion of a coliform colony from the confirmed test that is positive is transferred to a growth medium and inoculated for __________, and a second portion is transferred to __________ and incubated for __________.

    a. 18–24 hr, lauryl tryptose broth (LTB), 24–48 hr
    b. 24–48 hr, LTB, 24–48 hr
    c. 18–24 hr, brilliant green lactose (BGL), 24 hr
    d. 24 hr, BGL, 24–48 hr

40. Which device would best measure total suspended solids?

    a. Separatory funnel
    b. Büchner funnel
    c. Filter funnel
    d. Gooch crucible

41. If a radionuclide is above the MCL in finished water, how often must sampling be done and for how long?

    a. Monthly for at least 1 year
    b. Monthly until the running annual average is below the MCL
    c. Quarterly for at least 1 year
    d. Quarterly until the running annual average is below the MCL

42. __________ pipettes are used for performance monitoring and should be __________.

    a. Serological, Class 1
    b. Mohr, NSF approved
    c. Volumetric, Class A
    d. Büchner, Class 1

43. What piece of laboratory equipment can be made of polycarbonate and is clear, such that there is no confusion over reading the meniscus?

a. Kjeldahl flask
b. Volumetric flask
c. Graduated cylinder
d. Separatory funnel

44. Methyl orange used as an indicator in titrations is __________ on the alkaline side and __________ on the acid side.

a. colorless, orange
b. orange, red
c. yellow, salmon pink
d. reddish-orange, reddish-purple

45. If the multiple-tube fermentation or presence–absence methods are being used, as the presumptive positive samples are being inoculated into the __________, 0.1 mL of the presumptive broth is also transferred into __________ tube.

a. lauryl tryptose broth (LTB), an EC broth
b. LTB, a brilliant green broth (BGB)
c. BGB, an EC broth
d. BGB, an LTB

46. Separate refrigerators should also be used for the storage of __________ and __________ because the fumes of some chemicals may be harmful to __________.

a. chemical solutions, food, people
b. microbiological samples, reagents, the organisms
c. food, chemical solutions, people
d. microbiological samples, acid solutions, the organisms

47. A treatment plant's effluent water quality would be best shown from a sample collected _________.

a. on a major transmission main
b. on a lateral or minor pipe at some of the furthest points in the distribution system
c. immediately past the customer's meter
d. at a customer's tap, preferably a kitchen or bathroom sink

48. The analysis of a water sample yielded a result of 4.0 mg/L $CO_2$. What is the most likely source of this water sample?

a. Shallow well
b. Deep well
c. Surface water
d. A brackish lagoon

49. The dosage for lime and soda ash at a softening plant is based on _________.

a. calcium carbonate alkalinity
b. hydroxide and carbonate alkalinity
c. hydroxide, carbonate, and bicarbonate alkalinity
d. calcium and magnesium carbonate alkalinity

50. What results are used to develop the Langelier saturation index?

    a. Alkalinity, pH, temperature, $CaCO_3$ and $MgCO_3$ saturation, and $CO_2$
    b. Alkalinity, pH, $CaCO_3$ saturation, $MgCO_3$, and $CO_2$
    c. Alkalinity, pH, temperature, $CaCO_3$, and total dissolved solids
    d. Alkalinity, pH, $CaCO_3$ saturation, and $CO_2$

51. What is the most common radionuclide anion?

    a. Uranium
    b. Radium 226
    c. Radium 228
    d. Thorium

52. _________ and _________ should not be stored in the same refrigerator.

    a. Drinking water samples, food
    b. Drinking water samples, food and drinks
    c. Chemical solutions, samples
    d. Chemical solutions, reagents

53. Besides using carbon dioxide, encrustation can also be prevented by adding _________.

    a. sodium hexametaphosphate
    b. sequestering agents
    c. citric acid
    d. muriatic acid

54. In concentration cell corrosion, what gas is produced?

    a. Hydrogen
    b. Oxygen
    c. Carbon dioxide
    d. Ammonia

55. In the EC-MUG test, if the tube _________, it is positive for *E. coli*.

    a. fluoresces under a long-wave UV light
    b. fluoresces under a short-wave UV light
    c. turns yellow
    d. produces gas

56. The process of removing organics from a water sample and concentrating them in a solvent suitable for injection into the _________ is called _________.

    a. atomic absorption, liquid-phase microextraction
    b. atomic absorption, multiphase extraction
    c. mass spectrophotometer, solid-phase microextraction
    d. mass spectrophotometer, in situ microextraction

57. Color in the _________ range is determined by a visual comparison of the sample with either a known colored chemical solution or a calibrated color disk.

    a. greenish-gray
    b. green to brown
    c. yellow to brown
    d. orange to brown

58. The specific ultraviolet absorbance (SUVA) is calculated using the absorbance of UV light at 254 nm _________ the concentration of _________.

   a. multiplied by, TOC
   b. divided by, TOC
   c. multiplied by, dissolved organic carbon
   d. divided by, dissolved organic carbon

59. What type of oven would be used to burn solids?

   a. Muffle oven
   b. Autoclave
   c. Commercial oven
   d. Gravity convection

60. What is the log removal for a water treatment plant if the samples show a raw water coliform count of 150/100 mL and the finished water shows 1.2/100 mL?

   a. 2.0 logs removed
   b. 2.1 logs removed
   c. 2.2 logs removed
   d. 2.3 logs removed

61. A water that is neither scale-forming nor corrosive is stable. This condition is termed _________.

   a. saturation
   b. zero potential
   c. stabilization
   d. the Langelier zero index

62. What is the water chemist interested in with regard to pH?

   a. What is in the water, such as inorganics, organics, and algae, because they can influence the pH
   b. Whether the water is scale-forming or corrosive
   c. What makes up the pH and how much buffering exists
   d. The contaminants and their influence on pH

63. To determine the carbon dioxide concentration in water using the nomographic method, an operator also would have to know _________.

   a. bicarbonate alkalinity, total dissolved solids, and pH
   b. total dissolved solids and pH
   c. total dissolved solids, temperature, and pH
   d. temperature, pH, bicarbonate alkalinity, and total dissolved solids

64. When iron corrodes, it is converted from _________.

   a. $Fe$ to $Fe^{+2}$
   b. $Fe^{+2}$ to $Fe^{+3}$
   c. $Fe^{+3}$ to $Fe^{+2}$
   d. $Fe$ to $Fe^{+3}$

65. The readouts on electrical conductivity meters is in _________.

    a. microamps, 20°C
    b. microamps, 25°C
    c. microhms, 20°C
    d. microhms, 25°C

66. Regarding the USEPA sampling frequency for radionuclides in drinking water for individual contaminants at each entry point: If results of testing are equal to or above the detection limit but equal to or less than one-half the MCL, what is the frequency of sampling?

    a. 3 years
    b. 6 years
    c. 9 years
    d. 12 years

67. A simple test to determine the possible presence of pathogens would be testing for ________.

    a. *Giardia*
    b. *Cryptosporidium*
    c. coliforms
    d. *E. coli*

68. What is the HPC method, and how long are the samples incubated?

    a. A 10:1 dilution of water to sample is inoculated to an eosine methylene blue agar plate, 24–48 hr
    b. A 10:1 dilution of water to sample is inoculated to an eosine methylene blue agar plate, 48–72 hr
    c. Diluted water samples are placed on a plate-count agar, 24–48 hr
    d. Diluted water samples are placed on a plate-count agar, 48–72 hr

69. What indicates whether the small amount of fluid in the tip of a pipette that remains after it has drained should be blown out?

    a. A color band near the top of the pipette
    b. A frosted band near the top of the pipette
    c. A capital BO near the top of the pipette
    d. Absence of TD near the top of the pipette

70. The mass spectrophotometer has the ability to identify many organic compounds by their _________.

    a. mass and fragmentation pattern
    b. frequency of peaks and curve width patterns
    c. frequency and the area under curve(s) on a chromatograph chart
    d. position of peaks in time and area under curve(s) on a chromatograph chart

71. What type of chlorine-measuring method uses a device that contains a chlorine gas-permeable membrane?

    a. Osmotic electrode method
    b. Functionalized polysulfonic extraction method
    c. Membrane extraction method
    d. Polarographic method

72. It is recommended that operators using particle counters track particles in the size range of __________ because they are a possible surrogate for __________.

a. 3–5 µm, *Giardia*
b. 5–8 µm, *Giardia*
c. 3–5 µm, *Cryptosporidium*
d. 5–8 µm, *Cryptosporidium*

73. What colors are not assigned numerical color units?

a. Greenish-blue and red
b. Pink and red
c. Blue and red
d. Blue and gray

74. The radioactive chemical radium is chemically similar to __________.

a. potassium
b. iron
c. calcium
d. phosphorus

75. The completed test is positive if gas is produced in the lauryl tryptose broth and __________ are found.

a. purple-stained, spore-forming, filamentous bacteria
b. red-stained, nonspore-forming, rod-shaped bacteria
c. pink-stained, spore-forming, cocci-shaped bacteria
d. red-stained, nonspore-forming, filamentous bacteria

76. Phenolphthalein appears __________ at a pH below 8.3 and __________ above pH 8.3.

a. pink, blue
b. red, blue
c. blue, red
d. colorless, pink or red

77. What would be the most economical and most common tool for evaluating of the coagulation process for a water treatment plant?

a. Streaming current monitor
b. Zeta potential measurement
c. Filterability tests
d. Jar tests

78. In the presence–absence (P-A) test, the presence of total coliforms is indicated by the __________ P-A medium turning __________ and by the formation of __________ in the medium.

a. purple, pink, gas
b. brown, pink, shiny pink colonies
c. purple, yellow, gas
d. brown, yellow, shiny yellow colonies

79. The SPADNS method is a procedure used to determine the concentration of ____________ in water. The SPADNS is the chemical ___________ in the test.

    a. fluoride ion, product
    b. fluoride ion, reagent
    c. arsenic ion, product
    d. arsenic ion, reagent

## Equipment Operation and Maintenance

1. When there are rapid pressure and velocity changes, some of the __________ comes off the pipe and causes "red water."

    a. FeOH
    b. $Fe(OH)_2$
    c. $Fe(OH)_3$
    d. $Fe_2O_3$

2. Electric energy is usually expressed in ___________, and power is expressed in ___________.

    a. amps, kilowatt-hours
    b. volts, amps
    c. kilowatt-hours, watts
    d. watts, amps

3. What is the total force in pounds exerted on the bottom of a tank with a diameter of 30.0 ft if the pressure at the bottom is 162 psig?

    a. 640 lb
    b. 113,000 lb
    c. 16,300,000 lb
    d. 4,069,500 lb

4. If a UV reactor is operating outside of the validation limits by more than ___________ of the water treated, it is considered to be off-specification.

    a. 1%
    b. 2%
    c. 5%
    d. 10%

5. Before installing a new shaft sleeve, it should be heated to __________.

    a. 212°F for 30 min
    b. 250°F for 15 min
    c. 212°F for 15 min
    d. 250°F for 30 min

6. At a ___________ and at a ___________ chlorine dioxide can revert to ___________.

    a. lower pH, low temperature, hypochlorite ion and sodium chlorite
    b. lower pH, high temperature, chlorine gas and chlorate
    c. higher pH, low temperature, hypochlorite and sodium chlorite
    d. higher pH, high temperature, chlorite and chlorate

7. Most ceramic-type diffusers in some aerators can be __________ trapped particles. Large bubble types can be cleaned with __________.

   a. heated to burn away, dilute citric acid solution and backwashing
   b. backwashed to remove, dilute citric acid solution and backwashing
   c. heated to burn away, a brush and detergent
   d. backwashed to remove, a brush and detergent

8. A condenser will allow __________ to flow through it, but it will block the flow of __________. One of the uses of condensers is to suppress __________.

   a. AC, DC, amperage spikes
   b. DC, AC, amperage spikes
   c. AC, DC, voltage surges
   d. DC, AC, voltage surges

9. The most used conductor material is _________.

   a. steel
   b. aluminum
   c. copper
   d. iron

10. Circuit breakers, compared to __________, have a much __________. This means that they are capable of interrupting the flow of considerably __________ than could be interrupted by __________.

   a. fuses, higher interrupting capacity, higher voltage, fuses
   b. switches, lower interrupting capacity, higher current, switches
   c. fuses, lower interrupting capacity, higher voltage, fuses
   d. switches, higher interrupting capacity, higher current, switches

11. How much stronger in disinfecting power is free chlorine compared to combined chlorine?

   a. 25 times
   b. 50 times
   c. 100 times
   d. 181 times

12. Why is ozone treatment often followed by granular activated carbon, which removes assimilated organic carbon caused by the action of the ozone treatment?

   a. To remove tastes and odors
   b. To remove color
   c. To prevent the regrowth of bacteria in the distribution system
   d. To prevent nitrification in the distribution system

13. The best packing for medium pressures and temperatures is _________.

   a. aluminum
   b. asbestos
   c. plant fiber
   d. Teflon

14. Excessive up-thrust on line shaft turbines is (are) caused by _________.
    a. misalignment of shaft
    b. pumping too much water
    c. too small or worn impeller
    d. vertical clearance between bottom of pump bowl and impeller is too small

15. Lubrication used on the bearings of line shaft turbines can be _________.
    a. oil
    b. grease
    c. water
    d. Either a or c

16. When do bar or wire-mesh screens usually encounter the largest amount of debris?
    a. Beginning of spring
    b. During spring
    c. Autumn
    d. Winter

17. What type of pump and impeller should be used for pumping sludge?
    a. Centrifugal pump with semi-open impellers
    b. Centrifugal pump with open impellers
    c. Reciprocating pump with closed impellers
    d. Reciprocating pump with semi-open impellers

18. The suction side of a pump's piping should be _________ pipe size _________ than the intake side of the pump.
    a. 0.5, smaller
    b. 1, smaller
    c. 0.5, larger
    d. 1, larger

19. Slimes on walls and appurtenances of sedimentation basins should be power-blasted off with fire hoses using pressurized clean water because they could harbor pathogens, such as _________.
    a. *Serratia*
    b. *Legionella*
    c. *Bacillus subtilus*
    d. *Naegleria gruberi*

20. Chlorine dioxide has gained popularity because it __________ than chlorine and is effective against __________.
    a. reacts with ammonia faster, all pathogens
    b. produces fewer disinfection byproducts, all pathogens
    c. reacts with ammonia faster, *Giardia*
    d. produces fewer disinfection byproducts, *Cryptosporidium*

21. In nanofiltration and reverse osmosis, membrane compaction rates increase when membranes are operated at __________ and __________.

    a. high flows, high pressures
    b. higher pressures, higher temperatures
    c. high pressures, the water contains high concentrations of iron
    d. high flows, the water has high organic concentrations

22. In a lead-acid battery, each of the battery cells produces an open circuit voltage of approximately _________.

    a. 1.8 V
    b. 2.0 V
    c. 2.1 V
    d. 2.7 V

23. Mixers that are immersed in fluoride solutions must be _________.

    a. made of carbon steel
    b. stainless steel
    c. coated with plastic material
    d. made of brass with zinc coating

24. What should be done to a shaft after an old sleeve is removed and before a new sleeve is installed?

    a. It should be thoroughly cleaned with a solvent and wiped down with a cotton cloth.
    b. It should be polished with emery cloth.
    c. It should be coated with a nonseizing compound, such as white lead.
    d. First do b, then c above.

25. Unlike reverse osmosis and nanofiltration, the recovery of the microfiltration process is not limited by the formation of __________ or the precipitation of __________.

    a. iron fouling, calcium carbonate
    b. silica fouling, carbonates
    c. scale, salts
    d. organic fouling, salts

26. An electric current in a wire travels from end to end at __________. However, individual electrons within the wire move __________ and over __________.

    a. high speeds, relatively slow, the entire length of the wire
    b. high speeds, relatively slow, short distances
    c. near the speed of light, at moderate speeds, the entire length of the wire
    d. near the speed of light, at moderate speeds, short distances

27. Garnet sand is used as the bottom medium in multimedia filters. What characteristics does garnet sand have compared to the other sand in the filter?

    a. Coarser and greater density
    b. Finer and greater density
    c. Smaller uniformity coefficient and slightly denser
    d. Coarser and less dense

28. Which of the following statements is false?
    a. Most pumps are self-priming.
    b. Bearings can be over lubricated.
    c. Fire point is defined as the point at which a lubricant will burn.
    d. Oil is used in high rpm conditions, and grease is used in low speeds and heavy loads.
29. What is used to coat the walls of sedimentation basins to rid the basin of slime growth?
    a. Sodium carbonate peroxyhydrate
    b. Dimethylalkylamine and lime
    c. Copper sulfate and lime
    d. Zinc pentahydrate
30. The viscosity of a fluid is most often measured in _________.
    a. SUS
    b. SAE
    c. SA
    d. SOA
31. The first stage of a pump gives 50 psi and 350 gpm. If two more stages are added, how much psi and gpm will it deliver?
    a. 50 psi, 1050 gpm
    b. 100 psi, 700 gpm
    c. 150 psi, 1050 gpm
    d. 150 psi, 350 gpm
32. Before reinstalling a shaft sleeve or impeller, it should be cleaned with _________.
    a. alcohol
    b. kerosene
    c. auto solvent
    d. Both b and c
33. Determine the cost to operate a 155 hp motor for one month (assume 30 days) if it runs an average of 6.77 h/d, is 84% efficient, and the electrical costs are $0.048 per kW.
    a. $946.89
    b. $1,701.47
    c. $31.56
    d. $1,269
34. A load, perpendicular to a pump shaft, applied to a bearing is called a _________.
    a. shear load
    b. thrust load
    c. radial load
    d. strain load
35. A pump with two stages produces a pressure of 60 psi. What would a third stage and a fourth stage produce?
    a. 120 and 240 psi
    b. 90 and 120 psi
    c. 90 and 270 psi
    d. 120 and 180 psi

36. What types of loads are applied to a pump's shaft?

a. Weight of rotating pump parts
b. Starting and running torque
c. Hydraulic forces
d. All of the above

37. How often should amperage and voltage tests be performed on three-phase motors?

a. Monthly
b. At least every 3 months
c. At least every 6 months
d. At least once per year

38. What should be used to clean an old bearing?

a. Solvent
b. Alcohol
c. Kerosene
d. A light oil

39. What should be done to a new shaft sleeve before it is put on a shaft if it uses an o-ring?

a. Heat the sleeve with a torch.
b. Heat the sleeve with oil.
c. Heat the sleeve with water.
d. Either b or c.

40. What chemical is used to raise alkalinity but requires the most maintenance?

a. Soda ash
b. Caustic soda
c. Sodium bicarbonate
d. Lime

41. An advantage of using ozone treatment is _________.

a. it improves coagulation
b. it is noncorrosive in small doses
c. its by-products are biodegradable
d. it is easy to feed

42. Because bromine is not as stable as chlorine, if it were used to treat public water supplies, how much more would be needed?

a. Two to three times more
b. Three to four times more
c. Four to five times more
d. Five to eight times more

43. What type of pump has the ability to rotate its discharge 360°?

a. End suction pump
b. Split case pump
c. Vertical pumps
d. Jet pumps

44. When chlorine dioxide is generated on site, a small amount of __________ can be formed.
    a. perchlorates
    b. perchloric acid and chloryl
    c. chlorate and chlorite
    d. dichlorine hexoxide
45. Microfiltration membranes made from PDVF have excellent chemical resistance to __________ and have a long service life ranging from __________.
    a. oxidants, 5–10 years
    b. acids and bases, 12–15 years
    c. organic fouling, 15–20 years
    d. iron fouling, 8–12 years
46. Cleaning protocols for microfiltration membranes vary according to the membrane tolerance to changes in __________ and the resistance to __________.
    a. pH, oxidation
    b. pressure, chlorine
    c. oxygen, chlorine
    d. temperature, oxidation
47. What term expresses electrical potential?
    a. Volts
    b. Amps
    c. Watts
    d. Ohms
48. Before installing casing wear rings, they should be _________.
    a. lubricated with a light oil
    b. greased with lithium
    c. cooled
    d. heated
49. Slow sand filters are cleaned by scraping the top __________ off the top of the sand filter.
    a. 1 in.
    b. 2 in.
    c. 3 in.
    d. 4 in.
50. Which of the following is not a part of a bearing?
    a. Rolling pin
    b. Inner and outer races
    c. Rolling element
    d. Cage
51. A short-term strategy for cleaning microfiltration systems for removing fouling species is _________.
    a. an acid treatment
    b. cleaning with surfactants
    c. reverse flow and air scrubbing
    d. acid treatment followed by surfactants

52. A soda ash tank is conical at the bottom and cylindrical at the top. If the diameter of the cylinder is 18.1 ft with a depth of 24 ft and the cone depth is 14.5 ft, what is the volume of the tank in cubic feet?

    a. 7,400 $ft^3$
    b. 6,175 $ft^3$
    c. 9,904 $ft^3$
    d. 29,676 $ft^3$

53. What is the correct order of parts that follows a pump for the listed components?

    a. Spool, concentric reducer, check valve, gate valve
    b. Concentric reducer, spool, check valve, gate valve
    c. Gate valve, spool, check valve, concentric reducer
    d. Spool, check valve, gate valve, concentric reducer

54. How should a pump's shaft be stored when not in use?

    a. Stored vertically, on end
    b. Stored horizontally in foam or other soft material
    c. Stored in a shaft box and coated with a light oil to prevent corrosion
    d. Stored horizontally in a long box on foam that is saturated with a light oil

55. Before installing impeller wear rings, they should be ________.

    a. lubricated with a light oil
    b. greased with lithium
    c. cooled
    d. heated



## Source Water Characteristics

1. Short-range planning for a water system's source of supply should be ________.

    a. 2–3 years
    b. 3–5 years
    c. 5–10 years
    d. 10–15 years

2. What increases when zebra mussels are present in great numbers in a body of water?

    a. Deposition of organic matter
    b. Aquatic insects feeding off dead decaying zebra mussels
    c. Seagulls
    d. pH

3. In the wellhead protection program, the most important technical step is ________.

    a. getting the funding for the research
    b. determining the zone of influence a particular well has
    c. determining the recharge area of a particular aquifer
    d. determining the extent of a particular aquifer and its watershed

4. Recommended sedimentation goals for source water that is 10 ntu or less is for the sedimentation effluent to be __________ of the time.

   a. <1 ntu 90%
   b. <1 ntu 95%
   c. <2 ntu 90%
   d. <2 ntu 95%

5. In a buttress dam, buttresses support reinforced concrete slabs and the impounded water's weight is used to help resist _________.

   a. lateral and vertical thrust as well as sliding at the base
   b. sliding and overturning by its own weight
   c. thrust from the water being held and stress at the base due the weight of the dam and the water
   d. vertical strain and lateral thrust

6. A gravity dam resists both _________.

   a. lateral thrust and vertical stress
   b. vertical thrust and lateral stress
   c. forward thrust from the water and strain from its own weight
   d. horizontal pressure of water and overturning by its own weight

7. A well is installed through the following formations: clay, medium sand, clay, silt, clay, coarse sand, clay, gravel, and finished in a zone of clay. What layer will contain the most water per cubic yard, and what layer will have the lowest pumping costs if each layer is similar in thickness?

   a. most water: gravel layer, lowest pumping costs: gravel
   b. most water: silt layer, lowest pumping costs: gravel
   c. most water: coarse sand, lowest pumping costs: medium sand
   d. most water: medium sand, lowest pumping costs: silt

8. Long-range planning for a water system's source of supply should be _________.

   a. 15–30 years
   b. 20–30 years
   c. 25–50 years
   d. 40–60 years

9. Superchlorination is typically used when __________ are excessive.

   a. heterotrophic plate counts
   b. quercitannic acids
   c. ammonia concentrations
   d. fulvic acids and humic acids

10. Hydraulic fracturing uses pressure to force__________ into existing fractures to open and separate different __________.

    a. air, mineral boundaries
    b. water or fluid, strata
    c. air, strata
    d. water or fluid, mineral boundaries

11. Ammonia is an inorganic compound occurring _________.
    a. predominately from wastewater discharges.
    b. predominately from domestic or wild animals.
    c. naturally or from domestic and wild animals.
    d. naturally or from wastewater discharge, domestic animals, and wild animals.
12. What are the principal precursors that react with disinfectants to produce disinfection by products?
    a. Volatile organic compounds and synthetic organic chemicals (SOCs)
    b. SOC
    c. Petroleum products
    d. Humic and fulvic acids
13. __________ or small streams at the upper end of a reservoir can help reduce __________.
    a. Installing a sediment trap, erosion
    b. Creating an artificial wetland, erosion
    c. Installing a sediment trap, siltation
    d. Creating an artificial wetland, siltation
14. Actinomycetes are a group of organisms usually associated with tastes and odors. This group has the characteristics of ________.
    a. fungi
    b. both bacteria and some algae
    c. both bacteria and fungi
    d. algae
15. In groundwater, the total dissolved solids will change __________ and the dissolved oxygen will change _________, with the dissolved oxygen lowest most probably around __________.
    a. slowly, rapidly, 4 p.m.
    b. rapidly, slowly, 4 p.m.
    c. seasonally, daily, early morning
    d. frequently, daily, late evening
16. Methane dissolved in water will cause ________.
    a. a musty taste
    b. a garlic-like taste
    c. a moldy taste
    d. absolutely no taste at all
17. What organism was introduced to the United States from Southeast Asia and has become a significant pest in almost every river system?
    a. Zebra mussels
    b. Grass carp
    c. Amus pike
    d. Asiatic clam *(Corbicula fluminea)*

18. Iron bacteria, such as __________, cause red water, tastes and odors, clogged pipes, and pump failure.

    a. *Yersinia*
    b. *Chlamodophila*
    c. *Gallionella*
    d. *Neisseria*

19. When temperature increases, the corrosion rate usually __________, and when the water is corrosive and the velocity increases, the corrosion rate __________.

    a. increases, increases
    b. decreases, decreases
    c. decreases, increases
    d. increases, decreases

20. Manganese is naturally found in soils primarily as __________ and in the two forms of __________.

    a. MnO, $Mn^{+2}$
    b. $MnO_2$, $Mn^{+2}$ and $Mn^{+4}$
    c. MnO, $Mn^{+4}$ and $Mn^{+6}$
    d. $MnO_2$, $Mn^{+}$ and $Mn^{+2}$

21. Microstrainers will not remove ________.

    a. eggs of small aquatic animals
    b. viruses
    c. bacteria
    d. Both b and c

22. Cryptosporidiosis is caused by the *Cryptosporidium's* ________.

    a. macrogametes
    b. oocysts
    c. cysts
    d. trophozoites

23. How often should routine source water samples be collected to determine the type and quantity of algae present before algae control procedures are implemented?

    a. Daily
    b. Weekly
    c. Monthly
    d. Quarterly

24. Synthetic organic compounds would more likely contain __________ than natural organic compounds.

    a. Phosphorous
    b. Sulfur
    c. Fluorine
    d. Oxygen

25. If a water system exceeds the 90th percentile of either the lead or copper action levels, it must also assess ________.

   a. its source water
   b. its pH levels in the distribution system and increase pH if too low
   c. its entire treatment process
   d. whether to implement a point-of-use program for the areas affected

26. Driven wells are __________; however, they are practical only when the water-bearing formations are __________ the surface and no __________ exist in the soil.

   a. relatively cost-effective to install, high yielding and close to, contaminants
   b. relatively cost-effective to install, within 80 ft of, contaminants
   c. simple to install, relatively close to, boulders
   d. simple to install, within 150 ft of, rocks

27. The depth to which submerged plants will grow is primarily limited by _______.

   a. climate
   b. depth of sunlight penetration
   c. amount of floating plants
   d. nutrient levels of the bottom mud

28. An earth dam is typically used where there is an abundance of __________ and where bedrock is __________. A concrete dam is often used where there is __________ and/or bedrock is __________.

   a. boulder-sized rock material, deep, scarcity of fill, deep
   b. fill material, shallow, an abundance of fill, shallow
   c. fill material, very deep, scarcity of fill, shallow
   d. rock material, shallow, an abundance of fill, deep

29. What may occur at the bottom of a lake when, in a cold climate, the upper layer is covered with ice and the bottom layer is warmer?

   a. It becomes oxygen-rich.
   b. The metabolism of aquatic animals slows way down.
   c. It becomes an oxidative environment.
   d. It becomes anaerobic.

30. When is the best time to start using a destratification equipment to overturn a lake?

   a. In spring when the average temperature of the air is higher than the water
   b. In summer when the average temperature of the air is higher than the water
   c. In fall to prevent stratification and thus anaerobic conditions from occurring
   d. Numerous studies have shown that it does not matter when destratification starts.

31. What primary problem will most likely occur if all aquatic plants are removed from a lake?

   a. Aesthetics of the lake may be ruined
   b. Food for waterfowl will disappear
   c. Algae problem may worsen
   d. May deplete invertebrates that fish feed on

32. What parameter(s), if it changes (they change) rapidly, would indicate a possible "dead zone" in a water body?
    a. pH and alkalinity
    b. Conductivity
    c. Dissolved oxygen
    d. Both a and c
33. *Cryptosporidium* and *Giardia* range in size from ________.
    a. 2 to 8 microns
    b. 3 to 15 microns
    c. 5 to 18 microns
    d. 7 to 20 microns

## Security, Safety, Compliance, and Administrative Procedures

1. What would be the first thing an operator should do if a moderate-to-large earthquake occurred near the water plant?
    a. Call the supervisor for direction
    b. Shut the plant down immediately
    c. Check all SCADA to make sure all processes are functioning
    d. Walk the plant and check for problems, starting with the most critical processes first.
2. What is the maximum permissible dose for radiation that nonradiation workers can be exposed to in rem/year?
    a. 0.5
    b. 0.8
    c. 1.0
    d. 1.2
3. What type of process control uses a mode with which the output of the controller is directly proportional to the deviation from the set point?
    a. Proportional control
    b. Floating proportional control
    c. Proportional plus reset control
    d. Proportional plus reset plus derivation control
4. If employers require employees to be chemical spill responders, then they have to give them initial training. Then, each year, they must have an annual refresher course that is __________ hours in duration.
    a. 4
    b. 8
    c. 12
    d. 16

5. States can grant a variance to a water system that cannot comply with an MCL __________. The variance may only be granted to systems that have __________ for treatment of the MCL being violated.

   a. because of characteristics of the water source, enhanced coagulation
   b. because of characteristics of the water source, installed full-scale best available technology (BAT)
   c. for a particular contaminant, enhanced coagulation
   d. for a particular contaminant, installed full-scale BAT

6. In the Lead and Copper Rule, if a water system exceeds the 90th percentile for the lead action level, the system must also _________.

   a. provide point-of-use devices to the customers in the affected area
   b. educate the affected public
   c. install or increase an existing process to raise the alkalinity in the finished water
   d. provide bottled water to the affected public until the problem is solved

7. The concentration of a disinfectant multiplied by the contact time from application to first customer must be calculated ________.

   a. continuously, using online instrumentation and SCADA
   b. daily
   c. weekly
   d. monthly

8. The Long-Term 2 Enhanced Surface Water Treatment Rule came into effect in 2006 for large systems. What tests were added to this rule that were not in the Surface Water Treatment Rule?

   a. *Cryptosporidium* and turbidity
   b. *Cryptosporidium*, *Giardia*, and turbidity
   c. *Cryptosporidium*, *E. coli*, and turbidity
   d. *Giardia* and *E. coli*

9. What is a disadvantage of using chlorine dioxide?

   a. It does not remove color.
   b. It adds a taste to the water that some people find objectionable.
   c. It does not oxidize organics very well.
   d. It is ineffective against lowering THM formation potential.

10. In 2003, the USEPA revised the drinking water standard for arsenic from ________.

   a. 10 µg/L to 1 µg/L
   b. 25 µg/L to 5 µg/L
   c. 50 µg/L to 10 µg/L
   d. 100 µg/L to 25 µg/L

11. What is a disadvantage of using caustic soda to soften hard water compared to other chemicals?

   a. More sludge
   b. Harder to store
   c. More expensive
   d. Harder to handle

12. What type of process control is limited because a correction can only be made after deviation has occurred and elements such as meters, transmitters, and transducers are needed?

    a. Proportional control
    b. Floating control
    c. Feedback control
    d. Two-position control

13. Isotopes are variations of the same __________ which have the same __________ but a different number of __________.

    a. element, atomic number, neutrons
    b. element, negative charges, neutrons
    c. compound, masses, protons
    d. molecule, charge, protons

14. If the initial scan for gross alpha particle activity is less than __________, the state may allow reduced monitoring frequency in the future.

    a. 1 pCi/L
    b. 2 pCi/L
    c. 3 pCi/L
    d. 5 pCi/L

15. Chlorite is regulated under the __________ and has a maximum contaminant level of __________.

    a. Stage 1 Disinfectants and Disinfection Byproducts Rule, 0.5 mg/L
    b. Stage 1 Disinfectants and Disinfection Byproducts Rule, 1.0 mg/L
    c. Stage 2 Disinfectants and Disinfection Byproducts Rule, 0.5 mg/L
    d. Stage 2 Disinfectants and Disinfection Byproducts Rule, 1.0 mg/L

16. Ammonia is a dangerous gas and is readily detectable at concentrations of ________.

    a. 5–15 ppm
    b. 10–30 ppm
    c. 20–50 ppm
    d. 100–120 ppm

17. What organization establishes regulations for new chemical contaminant testing?

    a. NSF
    b. NSF and NIOSH
    c. USEPA
    d. All state primacy departments, NSF, and USEPA

18. The Long-Term 2 Enhanced Surface Water Treatment Rule has specified that UV disinfection can provide at least _________.

    a. 1.5-log removal inactivation for *Cryptosporidium*
    b. 1.5-log removal inactivation for *Giardia*
    c. 2.0-log removal inactivation for *Cryptosporidium*
    d. 2.0-log removal inactivation for *Giardia*

19. When natural gas and air are used to make $CO_2$ for recarbonation, good ventilation is required because dangerous concentrations of __________ are also produced.

    a. hydrogen sulfide
    b. formaldehyde
    c. carbon monoxide
    d. methane

20. Any treatment changes for corrosion control need to be checked first with _________.

    a. hospitals and industries in the service area
    b. the local wastewater treatment authority
    c. the office of the state's drinking water program
    d. the USEPA

21. Small- and medium-sized utilities are deemed to have optimal corrosion control if they meet the lead and copper levels for ________.

    a. two consecutive sampling periods
    b. three consecutive sampling periods
    c. six consecutive sampling periods
    d. nine consecutive sampling periods

22. If a water system exceeds the action level for lead and copper, it must measure other contaminants as well. What parameter must be measured in the laboratory?

    a. Calcium
    b. Conductivity
    c. pH
    d. Temperature

23. Water systems can be exempt from Stage 1 and Stage 2 Disinfectants and Disinfection Byproducts Rule if the source water total organic carbon level is __________, calculated as a running annual average.

    a. <1.0 mg/L
    b. <2.0 mg/L
    c. <4.0 mg/L
    d. <5.0 mg/L

24. The Filter Backwash Recycle Rule is intended to help utilities minimize potential health risks associated with recycling backwash water particularly with respect to _________.

    a. *Giardia*
    b. *Cryptosporidium*
    c. *Giardia* and *Cryptosporidium*
    d. all pathogenic organisms

25. One of the key elements of the Long-Term 2 Enhanced Surface Water Treatment Rule was to use __________ monitoring results to classify surface water sources into one of __________ USEPA-defined risk levels called bins.

    a. *Cryptosporidium*, three
    b. *Giardia*, four
    c. *Giardia*, three
    d. *Cryptosporidium*, four

26. The __________ requires __________ to be collected before water enters the distribution system while avoiding areas where added chemicals may affect the results.

    a. Surface Water Treatment Rule, turbidity
    b. National Primary Drinking Water Standards, turbidity
    c. Stage 1 Disinfectants and Disinfection Byproducts Rule, residual chlorine
    d. Surface Water Treatment Rule, residual chlorine

27. All sedimentation basins should be equipped with __________.

    a. guardrails around the basin
    b. life rings or poles
    c. a roof
    d. Both a and b

28. The Information Collection Rule is a federal regulation requiring __________ water systems to collect special information to build up a database that will assist in the development of __________.

    a. all, monitoring regulations
    b. small, monitoring regulations
    c. large, monitoring and treatment regulations
    d. all, monitoring and treatment regulations

29. What type of softening process would probably not meet the Surface Water Treatment Rule requirements?

    a. Excess-lime softening
    b. Split treatment
    c. Caustic-soda treatment
    d. Lime–soda ash softening

30. Why is the bulk density shipping weight of a material different from laboratory density?

    a. Impurities in the shipping material
    b. Air mixed in with the shipping material
    c. Scales for shipping materials are less accurate
    d. Bulk weight uses trade weigh, which is 5% more

31. What was the primary concern for why the USEPA promulgated the Final Groundwater Rule?

    a. Fecal contamination
    b. Nitrites and nitrates
    c. Arsenic
    d. Lead and radon

32. A one-ton ammonia container has __________ fusible plugs and holds __________ of ammonia.

    a. two, 800 lb
    b. two, 950 lb
    c. three, 1,000 lb
    d. three, 1,200 lb

33. Under current regulations, what type of samples must be collected from the source water supply?

    a. Organic contaminants
    b. Radiological contaminants
    c. Coliforms
    d. *Cryptosporidium*

34. The Department of Health and Human Services and the USEPA have recommended that the optimum level of fluoride in drinking water is ________.

    a. 0.3 mg/L
    b. 0.5 mg/L
    c. 0.7 mg/L
    d. 1.0 mg/L

35. How does cyanide affect the human body?

    a. Causes kidney failure
    b. Causes nerve damage and respiratory problems
    c. Causes liver failure
    d. Causes nerve failure and kidney damage

36. What type of process control can be used if accuracy is not critical?

    a. Feedback control
    b. Feed forward control
    c. Floating proportional control
    d. Proportional control

37. What measuring and transmitting level device works readily on a 4–20 mA system but needs a temperature correction?

    a. Pressure activated
    b. Sonic type
    c. Float type
    d. Magnetic type

38. Groundwater systems at high risk for fecal contamination must provide a ________ inactivation of ________.

    a. 3.0-log, *Cryptosporidium*
    b. 4.0-log, *Cryptosporidium*
    c. 3.0-log, viruses
    d. 4.0-log, viruses

39. What is the Stage 1 Disinfectants and Disinfection Byproducts Rule maximum contaminant level for chlorite?

    a. 0.1 mg/L
    b. 0.5 mg/L
    c. 1.0 mg/L
    d. 2.0 mg/L

40. The Surface Water Treatment Rule requires that at no time should the water turbidity entering the distribution system exceed _________.

    a. 0.3 ntu
    b. 0.5 ntu
    c. 1.0 ntu
    d. 2.0 ntu

41. The maximum contaminant level for arsenic is ________.

    a. 5 µg/L
    b. 10 µg/L
    c. 15 µg/L
    d. 25 µg/L

42. __________ are a common-law doctrine, which means that they are part of a large body of __________ law.

    a. Riparian rights, civil
    b. Riparian rights, statutory
    c. Prior appropriations doctrine, civil
    d. Prior appropriations doctrine, statutory

43. Compared to Stage 1 Disinfectants and Disinfection Byproducts Rule (DBPR), what major change was made when Stage 2 DBPR was promulgated?

    a. The HAA5 maximum contaminant level was reduced to 60 mg/L.
    b. Each monitoring location has to comply with the running annual average.
    c. If total organic carbon substances (TOCs) are above 2.0 mg/L, the use of chlorine disinfectants at the beginning of the treatment process has to be reduced or eliminated.
    d. To remove specific percentages of TOC based on source water alkalinity.

44. If the TOC removal percentages cannot be met, a utility can comply with the EPA rule by removing at least __________ of _________, as $CaCO_3$, measured monthly and calculated quarterly as an annual running average.

    a. 5 mg/L, magnesium carbonate
    b. 10 mg/L, magnesium hardness
    c. 15 mg/L, magnesium hydroxide
    d. 20 mg/ L, magnesium carbonate

45. What organization issued the Risk Management Rule in 1996 in an effort to protect public health and safety (i.e., nonemployees)?

    a. OSHA
    b. USEPA
    c. The National Public Health Department
    d. The National Civil Rights Organization

46. This parameter is used to determine if the Disinfectants and Disinfection Byproducts Rule is in compliance.

    a. *Giardia*
    b. Total organic carbon
    c. Disinfectant residual
    d. Polyacrylamide

47. Under the _________, nonuse for a long period of time may result in loss of water right by forfeiture or abandonment.

  a. riparian doctrine
  b. prior appropriation doctrine
  c. absolute rights
  d. correlative rights

48. What is the maximum contaminant level for radon in drinking water?

  a. 0.2 pCi/L
  b. 0.5 pCi/L
  c. 1.0 pCi/L
  d. There is none.

49. Airline supplied air is not commonly used for emergency response because they are limited to only __________ ft.

  a. 100
  b. 200
  c. 250
  d. 300

50. Under the Emergency Response Planning Guidelines #2, what is the maximum airborne concentration of chlorine below which it is believed that nearly all individuals could be exposed for up to 1 hour without experiencing or developing irreversible or other serious health effects that could impair an individual's ability to take protective action?

  a. 3 ppm
  b. 4 ppm
  c. 5 ppm
  d. 10 ppm

51. The Radionuclides Rule added for the first time an MCL for uranium of _________.

  a. 5 μg/L
  b. 10 μg/L
  c. 25 μg/L
  d. 30 μg/L

52. In the Ground Water Rule for systems that disinfect, large systems are required to monitor __________ and small systems are required to monitor _________.

  a. every 4 hours, daily
  b. continuously, daily
  c. continuously, once per shift
  d. every 4 hours, once per shift

53. What genus type or organism has killed the most people because of its high mortality rate when contracted?

  a. *Shigella*
  b. *Vibrio*
  c. *Entamoeba*
  d. *Cryptosporidium*

54. What type of sludge from a drinking water plant cannot be discharged to a sanitary sewer system?

    a. Sludge containing acrylamide
    b. Ferric chloride
    c. ACH-containing sludge
    d. Lime sludge

55. The Long-Term 2 Enhanced Surface Water Treatment Rule also calls for continuous monitoring at the entry point to the distribution system with a minimum of __________ of chlorine residual.

    a. a trace
    b. 0.2 mg/L
    c. 0.5 mg/L
    d. 1.0 mg/L

56. What is under the state-level aquifer protection standard setting?

    a. Ways to secure funds for contaminated water cleanup
    b. It involves the evaluation of groundwater to determine the degree of contamination.
    c. It is used to impose regulations so that pesticides, herbicides, and fertilizers are used judiciously.
    d. It oversees the allocation and use of water with the goal of preventing saltwater intrusion and chemical contamination.

57. What disinfectant is effective in inactivating *Cryptosporidium* but has a high cost?

    a. Ozone
    b. Chloramine
    c. Chlorine dioxide
    d. Bleach

58. If a water system exceeds the action level for lead and copper, it must measure other contaminants as well. What parameter must be measured in the field?

    a. Conductivity
    b. pH
    c. Calcium
    d. Alkalinity

59. Some systems using aeration towers are required to install air scrubbers. These scrubbers use _________.

    a. sodium hydroxide
    b. granular activated carbon
    c. hydrated lime
    d. sodium bicarbonate

60. Emergency response at a water treatment facility generally means responding to a __________. This response is governed under the __________ and under __________.

    a. chemical release, Risk Management Rule, OSHA
    b. chemical release, Risk Management Rule, USEPA
    c. hazardous situation, Process Safety Rule, OSHA
    d. hazardous situation, Process Safety Rule, USEPA

61. A low-cost radioactive scan is first done to determine if the level of radioactivity is high enough to progress to further analyses. The level of radioactivity that would call for further analysis is normally the __________ for __________ or __________ but can be varied by the particular state drinking water agency.

a. MCL, uranium, thorium
b. detection limit, gross alpha, gross beta emitters
c. MCL, roentgen equivalent man (rem), radiation absorption dose (rad)
d. detection limit, uranium, thorium

62. In the Stage 2 Disinfectants and Disinfection Byproducts Rule, the new sampling sites are based on ________.

a. worst case sites throughout the distribution system
b. sites at the furthest points in the distribution system
c. lifetime distribution system evaluation
d. running annual average evaluations

63. What may be a sign that a child has been exposed to high lead levels?

a. Hemophilia
b. Myeloma
c. Nephritis
d. Anemia

64. Because microorganisms vary greatly in size, shape, and __________, one treatment method is seldom sufficient to inactivate or kill all of them.

a. surface charge
b. surface texture
c. chemical sensitivities
d. All of the above

65. How often must a system using ozone for oxidation or disinfection monitor the distribution system entry point for bromate?

a. Monthly
b. Quarterly
c. Biannually
d. Annually

66. Breathing $H_2S$ gas at concentrations as low as __________ by volume in air for less than __________ min can be fatal.

a. 0.01%, 20
b. 0.1%, 30
c. 0.5%, 20
d. 1.0%, 30

67. After the testing required in the Long-Term 2 Surface Water Treatment Rule is done, a system is put into a __________, which are __________ concentrations in the source water for the theoretical potential for contamination.

a. pack, minimum
b. pack, average
c. bin, minimum
d. bin, average

68. The Surface Water Treatment Rule requires that the turbidity of the water entering the distribution system is equal to or less than __________ in at least __________ of the measurements taken each month.

    a. 0.3 ntu, 90%
    b. 0.5 ntu, 90%
    c. 0.3 ntu, 95%
    d. 0.5 ntu, 95%

69. Drinking water containing copper that is well above the action level will most likely cause ________.

    a. heart palpitations
    b. nausea and vomiting
    c. neuropathy
    d. severe headaches

70. The Lead and Copper Rule is different from the other rules in that it deals with __________ to determine compliance.

    a. a treatment technique
    b. a treatment technique as well as an MCL
    c. point-of-use devices in certain circumstances
    d. a calculated percentile level

71. What NSF/ANSI Standard covers coatings, construction materials, and components?

    a. Standard 60
    b. Standard 61
    c. Standard 62
    d. Standard 63

72. What regulation requires performance goals for filters, data collection, and reporting exceptions for any filter not performing in accordance with the turbidity values that are stated in the rule?

    a. Interim Enhanced Surface Water Treatment Rule
    b. Surface Water Treatment Rule
    c. Filter Backwash Recycle Rule
    d. Total Coliform Rule

73. The rationale for dividing public water systems into three groups is the __________ of persons using the water.

    a. pathogen exposure
    b. possible cancer contaminant exposure
    c. nitrate and nitrite exposure
    d. chemical exposure

74. What are the credits given under the Interim Enhanced Surface Water Treatment Rule for *Giardia* that is using conventional treatment and with which some of the credit(s) are required to be achieved by disinfection with free chlorine, chloramines, ozone, or chlorine dioxide?

    a. 4.0 log
    b. 3.0 log
    c. 2.5 log
    d. 2.0 log

# III. Water Distribution, Introductory Questions

## Distribution System Components

1. What type of mechanical pipe cleaning unit has high-carbon spring-steel blades?

   a. Scraping pigs
   b. Metal scrappers
   c. Spiral pigs
   d. Bladed pigs

2. What will the loss of pressure in a water system possibly cause besides compromising firefighting capabilities?

   a. Pipe collapses
   b. Bacterial contamination
   c. Corrosion
   d. Collapsing pipes

3. __________ valves are most often used where pressure is relatively high and positive shutoff is required, such as the discharge of a high-lift pump.

   a. Large plug
   b. Gate
   c. Large globe
   d. Altitude

4. It is best to have an __________ to prevent the collapse of a pipe when it is being emptied.

   a. air-purging valve
   b. air-and-vacuum relief valve
   c. actuator
   d. atmospheric vacuum breaker

5. __________ have a major impact on the reliability of the distribution system to deliver an uninterrupted supply.

   a. Main break frequencies
   b. Breakdown of distribution pumps
   c. Power outages
   d. All of the above

6. What backflow prevention device is, at a minimum, required at the potable meter for domestic water booster pumps?
   a. Reduced-pressure zone backflow preventer
   b. Atmospheric pressure breaker
   c. Pressure vacuum breaker
   d. Double check valve
7. What is the purpose of a check valve?
   a. To control the rate of flow
   b. To control the velocity of flow
   c. To control the pressure
   d. To control the direction of flow
8. It is best to use __________ to determine the hydraulic grade line.
   a. piezometer readings
   b. pressure gauges
   c. elevations
   d. flow rate and elevations
9. Where would be the best place for hydrants to be located?
   a. Close to buildings
   b. Near the street
   c. Near street intersections
   d. Behind protective barriers
10. The most common meter register records the flow in ___________.
    a. imperial gallons
    b. cubic meters and gallons
    c. cubic meters
    d. cubic feet or gallons
11. This type of meter sends an electrical signal to a remote register mounted outside the building after a particular amount of water is used.
    a. Automatic meter reader
    b. Electrical pulse generator type
    c. Radio or satellite
    d. Scanning unit
12. How are distribution pump operations controlled?
    a. Water flow
    b. System pressure
    c. Control switches
    d. Altitude valves
13. The main distribution system performance categories that are included in the distribution system operator objective are ________.
    a. water quality and reliability
    b. water quality and power availability
    c. operating costs and emergency procedures
    d. operating costs and distribution capacity

14. What should an operator inspect first if a reduced pressure zone backflow preventer leaks at the vent?

 a. Inspect the first check valve
 b. Inspect the second check valve
 c. Check the regulator
 d. Check the test cocks

15. Which of the following factors are key for system design and important to consider when expanding distribution systems?

 a. Pressure
 b. Flow
 c. Water quality
 d. All of the above

16. Why should gaskets for piping not to be stored next to electrical motors?

 a. Excessive heat can ruin them.
 b. Oil from motor can damage them.
 c. Grease from motor can damage them.
 d. All of the above.

17. An 8.0-in.-diameter pipeline needs to be flushed. If the desired length of pipeline to be flushed is 145 ft, how many minutes will it take to flush the line three times at 45 gpm?

 a. 8.4 min
 b. 25 min
 c. 37.8 min
 d. 38 min

18. Auxiliary water sources regarding cross-connections can fall under the supervision of ________.

 a. OSHA
 b. a primacy agency
 c. the utility
 d. the local health department

19. It has been documented that _________ can pass through the walls of plastic pipe even though the pipe is carrying water under pressure. This process by which molecules pass through the plastic pipe is called _________.

 a. organic compounds, permeation
 b. organic compounds, infiltration
 c. acid compounds, permeation
 d. acid compounds, infiltration

20. The potential hazard for this cross-connection system is high.

 a. Swimming pools
 b. Boilers
 c. Commercial food processors
 d. Dishwashers

21. The dry-barrel hydrant should be completely opened when in use to ________.

 a. prevent the hydrant from vibrating and damaging internal components

b. ensure the drain valve is completely closed
c. prevent damage to the main valve due to throttling
d. ensure the operating screw does not get jammed

22. To prevent leakage at the point where the shaft protrudes through the case, either __________ are used to seal the space between these two.

a. packing rings or mechanical seals
b. lantern rings or mechanical seals
c. shaft sleeves or lantern rings
d. lantern rings or packing rings

23. Double check valves should be installed in areas that __________ as the __________ can form a cross-connection if submerged.

a. have a low risk for a biological hazard, relief port
b. have been approved by the primacy agency, relief port
c. have a low risk for biological hazard, test cocks
d. do not flood, test cocks

24. Why should a meter be placed as close to the property line as possible?

a. To prevent injury
b. To prevent it from being blocked
c. To prevent an illegal connection
d. All of the above

25. What will high water pressure cause?

a. More main and service leaks
b. More hot water tank failures
c. More appliance failures
d. All of the above

26. A meter being tested reads 3,685 L. A volumetric tank shows the actual value is 3,788 L. What is the percentage accuracy of the meter?

a. 102.8% meter accuracy
b. 51.4% meter accuracy
c. 95% meter accuracy
d. 97.28% meter accuracy

27. What restricts the flow between the impeller discharge and suction areas in a centrifugal pump to prevent excessive water circulation between the two?

a. Packing rings
b. Wear rings
c. Pump sleeves
d. Lantern rings

28. All hydrants should be inspected at least every _________.

a. 6 months
b. year
c. 2 years
d. 3 years

29. On a pump curve, what is plotted on the horizontal line and the vertical line?

    a. horizontal: capacity; vertical: head, power, and efficiency
    b. horizontal: capacity and power; vertical: head and efficiency
    c. horizontal: head; vertical: capacity, power, and efficiency
    d. horizontal: head and capacity; vertical: power and efficiency

30. An advantage of __________ is that there is no flow of water from a broken hydrant.

    a. warm-climate hydrants
    b. flush hydrants
    c. dry-barrel hydrants
    d. wet-barrel hydrants

31. What agency or regulation determines the frequency of testing and maintenance for a double check valve assembly?

    a. USEPA
    b. SDWA
    c. State and local codes
    d. ANSI/NSF Standard 63

32. What utility line will most likely not be moved for a water utility installing a new water main?

    a. Gas
    b. Sewer
    c. Electrical
    d. Telephone

33. Large motors are usually operated at _________.

    a. 400/420 V
    b. 410/430 V
    c. 440/480 V
    d. 470/490 V

34. A velocity-type flowmeter would be _________.

    a. magnetic
    b. Venturi
    c. flow tube
    d. orifice plate

35. The lowest minimum pressure in distribution systems of __________ is a standard in many states and water system guidelines.

    a. 20 psi
    b. 25 psi
    c. 35 psi
    d. 40 psi

36. Convert 7.94 mgd to cubic feet per second.

    a. 1.2 $ft^3/s$
    b. 5.1 $ft^3/s$
    c. 12.3 $ft^3/s$
    d. 13.7 $ft^3/s$

37. The __________ Standard __________ covers materials that come in contact with potable water and other issues.

    a. AWWA, 60
    b. ANSI, 61
    c. NSF, 60
    d. ANSI/NSF, 61

38. The first type of cleaning pig to be sent through a main is called __________.

    a. cleaning pigs
    b. bare pigs
    c. scraping pigs
    d. soft pigs

39. Most backflow incidents that cause health problems involve ________.

    a. kidney problems
    b. liver problems
    c. gastrointestinal problems
    d. nerve tissue damage

40. A distribution pipe is 1.93 mi long. What is the volume of water in gallons if the pipe is 2.00 ft in diameter for a length of 1.46 mi and 18.0 in. for the remainder?

    a. 148,272 gal
    b. 312,201 gal
    c. 213,840 gal
    d. 202,915 gal

41. In all motors fitted with ball bearings, roller bearings, or both, the bearing housings are fitted with __________ to keep the grease in and the dirt out.

    a. glands
    b. seals
    c. sleeves
    d. packing

42. What type of valve should be used when the pressure to be regulated is relatively low and tight shutoff is not important?

    a. Plug valve
    b. Slide valve
    c. Butterfly valve
    d. Pinch valve

43. This type of flowmeter measures the velocity based on the difference between the flow's dynamic pressure and the static pressure.

    a. Proportional meter
    b. Pitot
    c. Orifice plate
    d. Turbine meter

44. How can water operators control water age in the distribution system?

    a. Monitor chlorine residual, pH, and temperature
    b. Operate storage facilities to enhance mixing
    c. Test for disinfection by products and HPCs
    d. Lower tank levels by treating less water

45. The measurement devices most commonly used for measuring large quantities of water are __________ flowmeters.

    a. magnetic
    b. positive-displacement
    c. compound
    d. differential-pressure

46. What is the location of a lantern ring on a pump?

    a. Pump shaft
    b. Stuffing box
    c. Impeller shaft
    d. Motor shaft

47. Venturi meters are accurate for __________, have __________ friction loss, require __________, and have long been used for measuring flows in __________.

    a. low to high flows, low, low maintenance, medium and large pipelines
    b. low to high flows, high, high maintenance, medium and large pipelines
    c. a certain range of flows, high, low maintenance, larger pipelines
    d. a certain range of flows, low, low maintenance, larger pipelines

48. If the jets in a multi jet meter become __________, the meter will __________. If the jet orifice becomes __________, the meter will __________.

    a. worn, over-register, clogged, under-register
    b. worn, under-register, clogged, over-register
    c. clogged, under-register, worn, over-register
    d. clogged, over-register, worn, under-register

49. Internal pipeline corrosion control is usually implemented to reduce lead, copper, and ________.

    a. iron
    b. cadmium
    c. zinc
    d. manganese

50. Determine the volume of water in gallons for the following distribution system:

    Distribution pipe "A": 489 ft in length and 2.0 ft in diameter

    Distribution pipe "B": 2,655 ft in length and 1.5 ft in diameter

    Storage tank: 91 ft in diameter and a water height of 27.02 ft

    a. 1,160,000 gal
    b. 1,360,000 gal
    c. 1,314,000 gal
    d. 47,000 gal

51. In a magnetic flowmeter, water passing through the magnetic field induces a small flow of electrified current that is ___________, which is converted to flow rate.

    a. proportional to velocity
    b. proportional to velocity squared
    c. indirectly proportional to velocity
    d. indirectly proportional to velocity squared

52. Why are bypass valves often included within large gate valves?

    a. To equalize the pressure on the two sides of the gates
    b. For temporary flow when repaired or replacing the gate valve
    c. To clean the seat before closing the main valve
    d. To open when only small quantities of water are needed

53. What external coating is usually used to protect ductile-iron pipe?

    a. Epoxy
    b. Bituminous
    c. Polyvinyl fluoride
    d. Polybutylene

54. Power is defined as _________.

    a. the amount of energy put into a system
    b. the rate of doing work
    c. the amount of force per unit of time
    d. force multiplied by work

55. __________ relief valves should be at __________ points and __________ valves should be at __________ points in the distribution system.

    a. Air, high, blowoff, dead end
    b. Air-and-vacuum, high, blowoff, low
    c. Air, regular interval, blowoff, dead end
    d. Air-and-vacuum, regular interval, backflow, low

56. Prestressed, steel cylinder piping is a type of _________.

    a. steel cylinder pipe
    b. reinforced concrete pressure pipe
    c. steel pipe
    d. noncylinder steel pipe

57. Large water mains that surround and contribute water supply within the grid from several different directions would be a(n) _________.

    a. grid system.
    b. grid and tree system.
    c. arterial-loop system.
    d. arterial-loop and grid.

58. What is the velocity of flow in feet per second for an 8.00-in.-diameter pipe if it delivers 385 gpm?

    a. 1.51 fps
    b. 2.46 fps
    c. 0.407 fps
    d. 0.66 fps

59. Over time, __________ can cause damage to the pipeline and equipment because it (they) occur(s) so frequently.

a. surges
b. pump startup
c. flow demand changes
d. controlled pump shutdowns

60. What is a pump design called if it has the impeller mounted on a separate shaft, which is connected to the motor with a coupling?

a. Close-coupled
b. Frame-mounted
c. Horizontal-case
d. Vertical-case

61. The amount of metal removed from a pipe wall is __________ to the magnitude of the current flowing in the corrosion cell.

a. proportional
b. inversely proportional
c. indirectly proportional
d. equal

62. Throttling a pump against a gate valve should be avoided because it will _________.

a. eventually damage the pump
b. damage the gate valve
c. reduce efficiency of the pump and thus increase costs
d. Both b and c

63. A water storage tank is partially underground. The tank is 49.5 ft in diameter and 18.0 ft high. What is the upward force of water pressure in pounds lifting up the tank if the groundwater is 7.15 ft above the base of the tank?

a. 4.30 lb
b. 446 lb
c. 1,920 lb
d. 858,000 lb

64. Under this allocation of groundwater, this rule requires sharing.

a. Correlative rights
b. Reasonable use
c. Riparian rights
d. Prior appropriation doctrine

65. Metals that get into the distribution system due to a cross-connection will cause

a. nervous system damage.
b. nervous system and liver damage.
c. vomiting and liver damage.
d. All the above.

66. Usually steel pipe is used for _________.
    a. plant piping
    b. the suction side of pumps
    c. long-distance transmission mains
    d. initial feed lines to tanks and pump stations

67. What important role do system operators play in a water utility's management pipeline rehabilitation and replacement program?
    a. Setting renewal priorities
    b. Decisions on implementing the plan
    c. Cost decisions
    d. Impact of the program

68. The determining factor in sizing mains, storage facilities, and pumping facilities for communities with a population of less than 50,000 is usually the need for _________.
    a. fire protection
    b. domestic and irrigation use
    c. commercial use
    d. industrial use

69. What valve, if closed too fast, would cause serious water hammer compared to the others listed?
    a. Plug valve
    b. Globe valve
    c. Butterfly valve
    d. Altitude valve

70. A 10.0-in. main line needs to be flushed. If one volume of a 310-ft section of the pipeline was flushed through a fire hydrant for 15 min, what was the flushing rate in gallons per minute?
    a. 1,213.5 gpm
    b. 607.2 gpm
    c. 84.2 gpm
    d. 42.1 gpm

71. A water storage tank with no interior coating has a cathodic protected system installed. What will happen to the anodes?
    a. The anodes will disintegrate the same as if there was an interior coating.
    b. The anodes will not disintegrate and will not protect the tank.
    c. The anodes will disintegrate slightly faster.
    d. The anodes will disintegrate quickly.

72. Convert 73.4% to decimal form.
    a. 0.73
    b. 0.734
    c. 7.34
    d. 73.4

73. A tank is 74.5 ft in diameter and is 42.5 ft to the overflow. If there are 1,375,000 gal of water in the tank, what is the psi at the bottom of the tank?

a. 17.4 psi
b. 18.3 psi
c. 97.5 psi
d. 316.6 psi

74. In extremely cold conditions, meter pits are sometimes constructed with __________ to provide added insulation.

a. a double cover
b. straw packed in the pit above the meter
c. straw and gravel to just above the meter
d. fiberglass insulation

75. A confined aquifer is a(n) __________ of rock that is __________ that is (are) relatively __________.

a. permeable layer, confined by an upper layer, semipermeable
b. permeable layer, confined by upper and lower layers, impermeable
c. impermeable layer, confined by an upper and lower layer, semipermeable
d. impermeable layer, confined by an upper layer, permeable

76. A 10.0-in.-diameter distribution pipe delivers 1,287,000 gal in exactly 24 hours. What is the average flow during the 24 hours in feet per second?

a. 0.59 fps
b. 1.99 fps
c. 0.46 fps
d. 0.502 fps

77. How are the metal parts on concrete pipe protected from corrosion?

a. Resin
b. Mortar
c. Bituminous
d. Creosote

78. The pipeline network design should strive to eliminate or at least reduce the number of _________.

a. pump stations
b. pressure relief stations
c. dead ends
d. storage tanks

79. What is an operator's primary defense against backflow from cross-connections?

a. Performing testing and maintenance of backflow-prevention devices
b. Maintaining adequate system pressure
c. Funding and a clearly and concisely defined service policy
d. Public knowledge and support for the problems associated with cross-connections

80. Sampling for this particular site could be continuous, daily or on a weekly basis.

a. Storage tank
b. Dead-end area
c. Pressure relief station
d. Pump station

81. A tachometer generator is a sensor for measuring the _________.

    a. rotational speed of a motor's rotor
    b. rotational speed of a shaft
    c. the power output in watts or kilowatts.
    d. the voltage, usually of three-phase motors

82. In a sewage treatment plant a(n) __________ is required to isolate an area inside the plant and a __________ at the potable meter.

    a. vacuum pressure breaker, double check valve
    b. air gap, reduced-pressure zone backflow preventer
    c. vacuum breaker, reduced-pressure zone backflow preventer
    d. air gap, double check valve

83. What type of measuring level system has to compensate for the air temperature because the speed of sound in air varies with temperature?

    a. Diaphragm elements
    b. Ultrasonic system
    c. Bellows
    d. Bourdon tube

84. What pipe material, when new, has a C value of 150?

    a. Polyvinyl chloride
    b. Mild steel
    c. Ductile-iron pipe
    d. Cast iron pipe

85. Surface water will most probably _________.

    a. be repumped to boost water pressure after treatment
    b. be of high turbidity when entering the distribution system
    c. need greater-than-average storage capacity
    d. require softening due to excessive hardness

86. What facilities require the highest level of cross-connection and backflow protection?

    a. Sewage plant and agricultural pesticide mixing tanks
    b. Hospitals and funeral homes
    c. Agricultural pesticide mixing tanks and hospitals
    d. All the above

87. Single-family residences are most commonly served with a __________ service line.

    a. $\frac{1}{2}$-in.
    b. $\frac{5}{8}$-in.
    c. $\frac{3}{4}$-in.
    d. 1-in.

88. A sonic meter that measures water velocity works by sending out sound pulses _________.

    a. toward the flow of the liquid, straight down the pipe
    b. alternately toward the flow, then away from the flow straight down the pipe
    c. alternately toward the flow, then away in a diagonal direction across the pipe
    d. simultaneously in both directions, straight down the pipe

89. Generally, __________, __________, and __________ pipes are ranked by utilities as the most favorable materials for maintaining water quality.

a. PTFE, PVC, concrete pressure
b. prestressed steel cylinder, ductile iron, cement-mortar-lined steel
c. PVC, concrete pressure, lined ductile iron
d. PTFE, ductile iron, cement-mortar-lined steel

90. Throttling the discharge valve to approximate system flow should be done when _________.

a. elevated storage is not available
b. it is an off-peak period
c. water usage is minimal
d. there is no check valve between the pump and discharge valve

91. All meters that operate on the principle of determining flow velocity are designed with the assumption __________ through the meter, and every manufacturer has a recommendation for the minimum straight-pipe distance that should be on __________ of a meter.

a. of laminar flow, the upstream side
b. that friction loss is minimized, either side
c. of laminar flow, either side
d. that friction loss is minimized, the upstream side

92. What type of backflow prevention device is not recommended as protection in situations in which a health hazard may result from failure?

a. Reduced-pressure zone backflow preventer
b. Double check valve
c. Air gap
d. None of the above

93. What type of distribution system piping generally runs in straight lines and has few side connections?

a. Transmission mains
b. Auxiliary lines
c. Treatment plant piping
d. Distribution mains

94. Valve vault drains should be connected to a(n) _________.

a. sanitary sewer
b. storm drain
c. absorption pit
d. nearby ditch or gutter

95. If a water tank has a volume of 3.25 MG and the flow from the tank is 6.10 mgd, what is the detention time in hours?

a. 0.787 hours
b. 0.02 hours
c. 12.8 hours
d. 1.28 hours

96. Pesticides that get into the distribution system due to a cross-connection will cause ________.

    a. nervous system damage
    b. flu-like symptoms
    c. vomiting and nausea
    d. All of the above

97. What is a disadvantage to using reinforced concrete pipe?

    a. Poor flow characteristics
    b. Needs special protection in high-chloride soils
    c. O-ring joints are difficult to install
    d. Requires extensive bedding and backfill preparation

98. How often should the anodes be inspected in a cathodic protected water storage tank?

    a. Every 6 months
    b. Annually
    c. 2–3 years
    d. 3–5 years

99. What can reduce the friction between the packing rings and the pump shaft?

    a. Polyethylene
    b. NSF-approved grease
    c. The water being pumped
    d. Graphite

100. If a meter is oversized, what could occur?

    a. The meter will over-register at high flows.
    b. The meter will under-register or not register anything at low flows.
    c. Nothing except cost the utility more for a meter that is too big.
    d. It will under-register at low flows and over-register at high flows.

101. One person should be able be to operate a hydrant using a ________ hydrant wrench, and if a longer one is needed, the hydrant should be __________.

    a. 12-in., repaired or replaced
    b. 14-in., dismantled to remove and replace the operating net and main stem
    c. 15-in., repaired or replaced
    d. 16-in., dismantled to remove and replace the operating net and main stem

102. The major disadvantage(s) of pits for water meters in cold climates is that __________.

    a. they often freeze
    b. snow can cover them, making them hard to find
    c. they can flood and thus need to be pumped out
    d. All of the above

103. What type of pipe, if dewatered rapidly, has the potential to collapse?

    a. CIP
    b. DIP
    c. Steel
    d. Unreinforced concrete

104. Meters that are 1½ and 2 in. most commonly use _________.

a. tapered meter couplings
b. regular pipe couplings
c. flanged meter couplings
d. a special meter coupling designed to provide a seat against a gasket

105. What type of pipe will use elastomeric gaskets?

a. Steel
b. Ductile iron
c. Concrete
d. High-density polyethylene

106. What is the problem with using ordinary comprehensive maps as basic layouts for a water utility?

a. They may not be accurate.
b. There is too much clutter that is not needed, making the utility updated map difficult to read.
c. The map symbols are all different and thus would need to be changed.
d. The cost savings is insignificant, thus it is best for the utility to make its own.

107. A pressure-reducing valve has two upper operating chambers sealed from each other by a _________.

a. float-operated needle valve
b. flexible, reinforced diaphragm
c. spring-loaded seat
d. throttling diaphragm

108. A cone valve has a movable internal part that is cone-shaped and is opened when the rotating __________ is turned through an angle of 90° so fluid can pass through a machined port.

a. plug
b. ball
c. cone
d. disk

109. What is the psi 15 ft above the bottom of a lake that is 367 ft deep?

a. 145 psi
b. 152 psi
c. 165 psi
d. 170 psi

110. Small motors are usually operated at _________.

a. 200/210 V
b. 220/240 V
c. 240/260 V
d. 190/200 V

111. What pipe is especially prone to sliding out of a push-on joint?

a. Ductile-iron pipe
b. Plastic encased pipeline
c. Asbestos-cement pipe
d. Unlined steel pipe

112. How much surface area is contained by a tank that has a radius (r) of 62.5 ft?

a. 196.25 $ft^2$
b. 12,300 $ft^2$
c. 24,500 $ft^2$
d. 24,532 $ft^2$

113. What can be used for relieving and preventing surges?

a. Check valves
b. Pressure-relief valves
c. Altitude valves
d. Air relief valves

114. Water storage tanks should be cleaned and inspected at least every

a. 1–2 years
b. 3–5 years
c. 6–8 years
d. 7–9 years

115. Find the volume in gallons of a trapezoidal trough that has the following dimensions:

Length = 82.1 ft

Height = 3.1 ft

Bottom width = 4.2 ft ($b_1$)

Top width = 7.9 ft ($b_2$)

a. 11,500 gal
b. 206 gal
c. 1,540 gal
d. 3,080 gal

116. Apparent water losses are caused by _________.

a. unknown leaks.
b. inaccurate meters.
c. unauthorized consumption.
d. All of the above.

117. Most control valves in the water system are intended to _________.

a. start and stop flows
b. regulate pressure and throttle flow
c. isolate piping
d. relieve pressure

118. Convert 4,994 gpm to cubic meters per second.

a. 3.15 $m^3/s$
b. 0.63 $m^3/s$
c. 0.315 $m^3/s$
d. 6.3 $m^3/s$

119. Water is flowing through a faucet at 13.9 gpm. How long will it take to fill a swimming pool in hours and minutes if the pool is 31 ft by 21 ft and averages 5.4 ft in depth?

a. 4 hours 14 min
b. 5 hours 50 min
c. 31 hours 32 min
d. 113,502 hours

120. Water is flowing at a velocity of 2.85 ft/s in an 8.0-in.-diameter pipe. If the pipe changes from the 8.0-in. to a 10.0-in. pipe, what will the velocity be in the 10.0-in. pipe?

a. 1.83 ft/s in 10.0-in. pipe
b. 14.98 ft/s in 10.0-in. pipe
c. 0.22 ft/s in 10.0-in. pipe
d. 4.45 ft/s in 10.0-in. pipe

121. What is the internal surface area of a cylindrical tank (bottom, top, and the cylinder wall), if it is 65.0 ft in diameter and 24.0 ft high? Assume the top is flat.

a. 4,898.40 $ft^2$
b. 6,633.25 $ft^2$
c. 11,531.65 $ft^2$
d. 23,062 $ft^2$

122. What type of pipe consists of one or more steel cages of welded wire fabric or helically wrapped rods welded to longitudinal rods?

a. Reinforced concrete noncylinder pipe
b. Reinforced concrete cylinder pipe
c. Bar-wrapped concrete cylinder pipe
d. Prestressed concrete cylinder pipe

123. What is effective height?

a. The total feet of head against which a pump must work
b. The distance between the centerline of a pump and the level it is pumping to, including friction and minor losses
c. The dynamic discharge head minus friction and minor losses
d. The static friction head

124. __________ is a measure of the ability of the material to bend without breaking.

a. Shear strength
b. Strain load
c. Flexural strength
d. Tensile strength

125. The plumbing within the customer's premises regarding cross-connections usually falls under the supervision of _________.

a. state and local health authorities
b. the utility
c. the USEPA
d. OSHA

126. The volume of water recorded on a turbine meter's register is __________ to the number of revolutions by the __________.

a. directly proportional, impeller
b. directly proportional, propeller
c. directly proportional, rotor
d. directly proportional, turbine

127. What is the average MG flow from a storage tank given the following data?

| Mon. | Tues. | Wed. | Thurs. | Fri. | Sat. | Sun. |
|---|---|---|---|---|---|---|
| 7.9 | 8.0 | 6.5 | 6.8 | 7.0 | 7.3 | 7.8 |

a. 6.3 mgd
b. 25.7 mgd
c. 7.3 mgd
d. 51.3 mgd

128. Pressure head can be described as __________.

a. the amount of energy that water possesses because of its elevation
b. the amount of energy due to velocity of flow
c. the vertical distance from point of pressure measurement to the hydraulic grade line
d. the gauge pressure due to elevation or velocity

129. Both __________ and __________ customers dislike day-to-day changes in hardness, tastes, temperature, or chemical composition.

a. commercial, industrial
b. commercial, domestic
c. industrial, domestic
d. chemical, industrial

130. If a tank has a circumference of 323 ft, what is the diameter?

a. 100 ft
b. 103 ft
c. 1010 ft
d. 1014 ft

131. Metal scrappers have on their body sections __________.

a. silicon carbide wire
b. high-carbon spring-steel blades
c. flame-hardened steel-wire blades
d. wire brushes imbedded into hardened polyurethanes

132. What type of pipe is the most difficult to tap and repair?

a. Concrete
b. Fiberglass
c. High-density polyethylene
d. Cross-linked polyethylene

133. According to best practices, what percentage of new meters should a water utility test?

a. 10%
b. 20%
c. 25%
d. 50%

134. What is the drawdown of a well if the well yields 129 gpm and the specific yield is 45 gpm/ft?

a. 0.35 ft
b. 13.7 ft
c. 287 ft
d. 5,805 ft

135. What type of distribution system piping in general has exposed piping?

a. Transmission mains
b. Storage tank facilities
c. Pump station piping
d. Distribution auxiliary piping

136. What is the heart of the information system for water utilities?

a. SCADA
b. PLCs
c. RTUs
d. Both b and c

137. What type of thrust control would be best on vertical bends?

a. Thrust anchors
b. Thrust blocks
c. Tie rods
d. Restraining fittings

138. The overflow pipe from an elevated tank should be brought down from the maximum tank level to a point about _________ off the ground surface.

a. 6 in.
b. 2 ft
c. 2.5 ft
d. 3 ft

139. After all packing rings are installed firmly, back off on the gland nuts about __________ before the pump is started.

a. 4 turns
b. 3 turns
c. 2 turns
d. 1 full turn

140. Because of the limited choice of fittings, a combination of special fittings must now be made for _________.

a. CIP
b. DIP
c. Asbestos-cement
d. Steel pipe

141. If a pump moves 435,000 gal in exactly 48 hours, how much will it move in seven hours if all other conditions remain the same?

   a. 9,060 gal pumped
   b. 63,400 gal pumped
   c. 31,700 gal pumped
   d. 18,500 gal pumped

142. What is used on the discharge side of a pump, starting next to the pump and going out?

   a. A globe valve, then a check valve
   b. A gate valve, then a relief valve
   c. A gate valve, then a check valve
   d. A check valve, then a gate valve

143. During an extremely severe winter, frost may penetrate __________ times deeper than the average winter frost depth.

   a. 1.5
   b. 2
   c. 2.5
   d. 3

144. One advantage of using mechanical seals, if operating properly, is they _________.

   a. are easy to replace
   b. do not have to remove the shaft and impeller from the case
   c. fail very slowly, so they can be changed before complete failure
   d. do not wear or damage shaft sleeves

145. A pressure relief valve is essentially just a __________ with an adjustable __________ to maintain pressure on the valve seat to keep the valve closed under normal pressure conditions.

   a. globe valve, spring
   b. globe valve, piston
   c. altitude valve, spring
   d. altitude valve, piston

146. Determine the head loss in psi for a pipe that has a water velocity of 2.95 fps.

   a. 0.312 psi
   b. 3.14 psi
   c. 0.058 psi
   d. 0.060 psi

147. The most popular pipe joint is the _________.

   a. ball-and-socket joint
   b. mechanical joint
   c. flanged joint
   d. push-on joint

148. A garden hose connected to a chemical dispenser would produce what type of potential cross-connection hazard?

a. No effect
b. Backpressure
c. Backsiphonage
d. Backpressure or backsiphonage

149. This type of pump uses closely meshed gears, vanes, or lobes.

a. Reciprocating pump
b. Rotary pump
c. Positive-displacement pump
d. Peristaltic pump

150. When testing positive-displacement meters for accuracy, how many different flows are usually used?

a. Two
b. Three
c. Four
d. Five

151. The relief valve on the discharge side of a pump is for _________.

a. protecting the pump from excessive surge pressures
b. flow control
c. pressure regulation
d. All of the above

152. Water quality design considerations for storage facilities must conform to the latest __________ and __________ standards.

a. AWWA, National Science Foundation (NSF)
b. AWWA, ANSI
c. USEPA, ANSI
d. USEPA, NSF

153. A water tank with a capacity of 2.85 MG is being filled at a rate of 1,750 gpm. How many hours will it take to fill the tank?

a. 299 hours
b. 27.1 hours
c. 0.04 hours
d. 29.9 hours

154. Turbine meters have __________, but water must be moving at sufficient speed before the _________ will start to rotate.

a. low friction loss, rotor
b. low friction loss, propeller
c. high friction loss, rotor
d. high friction loss, propeller

155. What type of pump has a relatively low initial cost but needs periodic checks to monitor impeller, packing, or mechanical seal condition?

    a. Radial-flow pump
    b. Mixed-flow pump
    c. Axial-flow pump
    d. Centrifugal pump

156. What type of pump is a multistage, mixed-flow, centrifugal pump or turbine pump with an integral or close-connected motor?

    a. Submersible pump
    b. Deep-well pump
    c. Booster pump
    d. Axial-flow pump

157. A trench that is 4.2 ft wide, 728 ft long, and 4.5 ft deep is excavated for the purpose of installing a water main. Determine the number of cubic feet and cubic yards that need to be excavated for this water main.

    a. 18.9 $yd^3$
    b. 510.3 $yd^3$
    c. 211 $yd^3$
    d. 510 $yd^3$

158. Each backflow prevention device should be tested for correct operation __________ by an individual specifically qualified for such testing and repair.

    a. quarterly
    b. semiannually
    c. annually
    d. every 2 years

159. A rectangular water storage tank is partially underground. The tank is 50.0 ft long, 24.0 ft wide, and 18.5 ft high. What is the upward force of water pressure in pounds lifting up the tank if the groundwater is 8.13 ft above the base of the tank?

    a. 2.4 lb
    b. 9,210 lb
    c. 1,057 lb
    d. 609,000 lb

160. A water tank with a capacity of 1.75 MG is being filled at a rate of 1,100 gpm. How many hours will it take to fill the tank?

    a. 26.5 hr
    b. 159 hr
    c. 10.5 hr
    d. 1,591 hr

161. In freezing climates, all hydrants should be inspected in the __________.

    a. winter
    b. early spring
    c. late summer
    d. autumn

162. Most couplings are of the __________ and require periodic maintenance, usually every __________.

a. lubrication style, 2–3 years
b. lubrication style, 6 months or at annual intervals
c. dry style, 6 months or at annual intervals
d. dry style, 2–3 years

163. Power systems in the United States operate at __________.

a. 60 cycles or 60 Hz
b. 60 cycles or 120 Hz
c. 30 cycles or 60 Hz
d. 60 cycles or 120 Hz

164. What is a serious disadvantage of using barometric loops?

a. They need high surveillance to make sure they are working properly.
b. Initial high cost
c. Too high a space requirement
d. Prevents only backpressure

165. A reduced-pressure zone backflow preventer must be included in locations where the relief port cannot ________.

a. be submerged
b. be vandalized
c. freeze
d. All of the above

166. A distribution main is defined as any pipe in the distribution system ________.

a. other than a service line
b. that is larger than 6 in
c. that is larger than 8 in
d. that is a branch for laterals that carry service lines

167. When testing current or compound meters for accuracy, how many flow rates are typically used?

a. 4
b. 6
c. 8
d. 10

168. In general, fire flow, when delivered to a fire department pumper, has a specified residual pressure of ________.

a. 20 psi
b. 25 psi
c. 35 psi
d. 40 psi

169. If the pressure head on a fire hydrant is 188 ft, what is the pressure in psi?

a. 434.3 psi
b. 81.4 psi
c. 418.0 psi
d. 88.3 psi

170. If a tank's radius is 49.8 ft, what is the area of the tank's bottom?

 a. 31,150 $ft^2$
 b. 313 $ft^2$
 c. 1,947 $ft^2$
 d. 7,790 $ft^2$

171. Fire protection for a community is based on the size of the _________.

 a. mains
 b. storage facilities
 c. pumping facilities
 d. All of the above

172. What type of booster pump is generally used in water distribution pumping?

 a. Reciprocating pump
 b. Positive displacement pump
 c. Vertical turbine pump
 d. Rotary pump

173. Water rights acquired by means of diverting and putting the water to beneficial use following procedures established by state statutes or courts would be _________.

 a. appropriative rights
 b. appropriation-permit system
 c. correlative rights
 d. riparian doctrine

174. If water flows out of a new well that taps an aquifer without the help of the pump, it is called a _________.

 a. spring
 b. artesian aquifer
 c. artesian well
 d. piezometric well

175. Customers that have a wide variation in water use should be supplied with a _________.

 a. detector-check meter
 b. compound meter
 c. magnetic meter
 d. Venturi meter

176. A 10.0-in. main line needs to be flushed. If the desired length of pipeline to be flushed is 1,625 ft, how many minutes will it take to flush the line at 50.0 gpm?

 a. 10,083 min
 b. 331,000 min
 c. 66 min
 d. 132 min

177. When a section of pipe is replaced, the new piece must have a pressure rating __________ of the piece being replaced.

 a. equal to that
 b. equal to or greater than that
 c. greater than that
 d. greater than 10%

178. What type of pump is commonly called a can pump?

a. Turbine pump
b. Jet pump
c. Mixed-flow pump
d. Double-suction pump

179. Ductile-iron pipe resembles what other type of pipe?

a. Cast-iron pipe
b. Noncylinder pipe
c. Steel pipe
d. Steel cylinder pipe

180. Determine the flow in fps if a water flow of 315 gpm is going through an 8.0-in. pipe.

a. 1.00 ft/s
b. 2.01 ft/s
c. 2.10 ft/s
d. 3.04 ft/s

181. If a tank is not tall enough to accept full system pressure without overflowing, __________ is installed. Also, flow out of this type of tank is usually __________.

a. a pressure-relief valve, restricted
b. an altitude valve, restricted
c. a pressure-relief valve, unrestricted
d. an altitude valve, unrestricted

182. Of the following, what type of pipe is still manufactured and used in the United States?

a. Cast iron
b. Ductile iron
c. Asbestos-cement
d. High-density polypropylene

183. Nitrates and nitrites that get into the distribution system due to a cross-connection will cause __________.

a. blue baby syndrome (methemoglobinemia)
b. kidney toxicity
c. liver damage
d. nervous system damage

184. When the reduced pressure zone backflow preventer is operating correctly and the __________ pressure exceeds the __________ pressure, the supply pressure opposes the __________ and keeps the valve closed.

a. supply, downstream, first check valve
b. downstream, supply, second check valve
c. supply, downstream, relief valve's spring tension
d. downstream, supply, relief valve's spring tension

## Equipment Installation, Operation, and Maintenance

1. When disinfecting a pipe, the chlorine solution is usually injected through a __________ at the point where the new main connects to the existing system.

    a. corporation stop
    b. hydrant
    c. blowoff
    d. air relief valve

2. When disinfecting storage facilities using the chlorinate-and-fill method, how long should the chlorinated water stay in contact with the storage facility before filling it all the way?

    a. 2 hr
    b. 4 hr
    c. 6 hr
    d. 8 hr

3. When using a trench box, trench width should be __________ greater than the outside diameter of the pipe.

    a. 0.5–1 ft
    b. 1–2 ft
    c. 2–3 ft
    d. 3–4 ft

4. If a hydrant is taken out of service for repairs, who must be notified?

    a. The residents or businesses served by this particular hydrant
    b. The local fire department
    c. The police department
    d. The county health department

5. If a well produces 107 gpm with a drawdown of 22.5 ft, what is the specific yield in gpm/ft?

    a. 4.76 gpm/ft
    b. 2.1 gpm/ft
    c. 4.86 gpm/ft
    d. 2.38 gpm/ft

6. What are the two principal types of rotary valves?

    a. Butterfly and gate
    b. Bypass and butterfly
    c. Bypass and gate
    d. Plug and ball

7. Gaskets can be ruined by ________.

    a. dirt, oil, and excessive heat
    b. excessive exposure to sunlight
    c. ozone
    d. All of the above

8. What type of hydrant has the barrel filled with water only when the main valve is open?

   a. Dry-barrel hydrants
   b. Wet-barrel hydrants
   c. Warm-climate hydrants
   d. Flush hydrants

9. A pipe resting on a rock may break if it is weak in ________.

   a. tensile and flexural strength
   b. shear and flexural strength
   c. tensile strength
   d. shear strength and strain load

10. If a water system's unaccounted-for water loss exceeds __________, a system-wide leak survey should be conducted.

    a. 10%
    b. 15%
    c. 20%
    d. 25%

11. A __________ seat ring is a ___________ ring mounted in the body of a __________, against which the moving disc of the valve closes.

    a. bronze, polished, pump
    b. stainless steel, polished, pump
    c. bronze, machined, hydrant or valve
    d. stainless steel, machined, hydrant or valve

12. In a new pipe installation, leakage is defined as the volume of water that must be added to the "full" pipeline to maintain a specific test pressure within a __________ range.

    a. 2.5-psi
    b. 5-psi
    c. 7.5-psi
    d. 10-psi

13. The operating nut on the top of a hydrant is __________-sided.

    a. three
    b. four
    c. five
    d. six

14. Flushing a newly installed pipe is done to ________.

    a. remove the high concentration of chlorine used for disinfection
    b. remove mud or soil and other debris
    c. remove entrained air
    d. All of the above

15. If the __________ is unchanged but the __________ is decreasing, it usually indicates that the __________.

    a. drawdown, static level, pump is losing efficiency
    b. static level, drawdown, pump is losing efficiency
    c. well yield, pumping water level, pump is running at too high an amperage
    d. drawdown, static level, pump is running at too high an amperage

16. Most water systems use fire hydrants that have two __________ nozzles and one __________ nozzle in which the larger nozzle is used to connect the _________.

    a. 2.5-in., 4-in., fire hose
    b. 2.5-in., 4.5-in., pumper suction hose
    c. 3-in., 4-in., fire hose
    d. 3-in., 4.5-in., pumper suction hose

17. Some lead pipe, as it gets old, will crack if flexed because it ________.

    a. polymerizes
    b. crystallizes
    c. undergoes chemical reduction
    d. undergoes chemical oxidation

18. Copper tubing used in interior plumbing is usually joined with __________, and copper that is buried is usually joined with __________.

    a. compression joints, solder joints
    b. flare joints, solder joints
    c. solder joints, flare and/or compression joints
    d. flare and/or solder joints, compression joints

19. How do operators contain the pressurized water in a wet tap?

    a. Needle valve
    b. Plug valve
    c. Flapper valve
    d. Gate valve

20. The most important purpose that backfill serves is to ________.

    a. distribute surface loads
    b. protect pipe from corrosion and other damage
    c. to keep the pipe in place
    d. to prevent the pipe from pulling apart or from thrust

21. What type of hydrants have the main valve controlling flow to all the outlet nozzles, and there is no drain mechanism?

    a. Flush hydrants
    b. Warm-climate hydrants
    c. Wet-barrel hydrants
    d. Corey hydrants

22. The valves on newer style curb stops are ________.

    a. ball valves
    b. gate valves
    c. plug valves
    d. poppet valves

23. Why is it best to perform a hydrant-flushing program at night besides causing less traffic disruption?

    a. Flushing at night will take less time
    b. Fewer customer complaints
    c. Because Storage levels will not drop anywhere like during the day
    d. Water demand is much less at night

24. If special trench bedding material is required for a pipe installation, it should be clean, __________, granular material up to __________ in size.

   a. uniformly graded, ½ in.
   b. well-graded, ½ in.
   c. uniformly graded, 1 in.
   d. well-graded, 1 in.

25. The transmission of sounds is best in __________ and __________, whereas they are the worst in __________ and __________.

   a. copper, asbestos, plastic, concrete
   b. cast iron, ductile iron, plastic, asbestos cement
   c. copper, steel, asbestos cement, concrete
   d. steel, ductile iron, plastic, concrete

26. What type of soil transmits the best noise?

   a. Sandy soils
   b. Loamy soils
   c. Clay soils
   d. Organic silty soils

27. On a 90° or 45° pipe bend, if shackle rods are used, they should be placed in _________ and be on the _________ of the bend.

   a. undisturbed soil, inside
   b. undisturbed soil, outside
   c. concrete blocks, inside
   d. concrete blocks, outside

28. Pipe __________ breakage may occur when a force exerted on a pipe exceeds the material's __________ or __________ strength.

   a. fatigue, strain, shear
   b. ductile, strain, tensile
   c. shear, tensile, flexural
   d. stress, shear, flexural

29. What kinds of methods are used to place grout in the annular space between the well face and the casing?

   a. Tremie pour and pumping
   b. Tremie pour and dump bailer
   c. Dump bailer, pumping, and water-pressure driving
   d. All of the above

30. The most common means of disinfecting water mains is the use of _________.

   a. liquid chlorine
   b. gas chlorine
   c. calcium hypochlorite
   d. sodium hypochlorite

31. Most corrosive characteristics and conditions affect the __________ of a soil.

    a. neutrality
    b. electrical resistivity
    c. alkalinity
    d. acidity

32. This type of valve reduces the differential pressure across the closed disk and makes the main valve easier to open and close.

    a. Inserting valve
    b. Pressure-reducing valve
    c. Relief valve
    d. Bypass valve

33. What is an advantage for installing a meter pit in areas with deep frost?

    a. The concrete box rarely floods.
    b. Heavy lid construction prevents vehicle damage.
    c. It eliminates entry into the building.
    d. It is easy to locate in the snow due to the metal lid.

34. The compaction of soil in a trench should be done on the __________ backfill if native soil is used and it is relatively __________.

    a. upper, coarse
    b. upper, dense
    c. lower, coarse
    d. lower, dense

35. If a fire hydrant is used to test a new pipeline installation for leakage and pressure, what extra piece of equipment is needed?

    a. Positive-displacement pump
    b. Accurate water meter
    c. Double check valve
    d. A second pressure gauge on hydrant

36. What method of dechlorination has the disadvantages of possibly overdosing, under dosing, and being hard to tell when the chemical has been depleted?

    a. Venturi injection system
    b. Spray feed systems
    c. Flow-through systems
    d. Gravity feed

37. In the slug method for disinfecting a pipe, the chlorine concentration of at least __________ should be used, and the flow rate should be such that the slug remains in contact with each point on the pipe for at least __________ as it passes through the main.

    a. 200 mg/L, 3 hr
    b. 200 mg/L, 4 hr
    c. 300 mg/L, 3 hr
    d. 300 mg/L, 4 hr

38. This type of pump generates high-velocity water and has relatively low cost and maintenance.

a. Reciprocating pump
b. Submersible pump
c. Centrifugal-jet pump
d. Deep-well pump

39. A well has a depth of 306.1 ft. If the depth to water is 124.7 ft, what is the pressure in psi at 5.8 ft above the bottom? Disregard the additional atmospheric pressure in the well.

a. 79 psi
b. 13 psi
c. 76 psi
d. 175 psi

40. The pumping water level in a well is 106.4 ft. If the drawdown was 14.2 ft, what must have been the static water level in the well?

a. 120.6 ft
b. 113.5 ft
c. 106.4 ft
d. 92.2 ft

41. Old pipes should be cleaned if they show signs of ________.

a. bacterial slime growth
b. deposits of iron bacteria
c. tuberculation
d. All of the above

42. What usually accounts for most of the energy used in water supply when gravity cannot be used?

a. Valving
b. Pumping
c. Lighting
d. Fuel for vehicles

43. If a trench is 112 ft long, 4.5 ft wide, and 5.9 ft deep, how many cubic yards of soil were excavated?

a. 84 $yd^3$
b. 110 $yd^3$
c. 2,974 $yd^3$
d. 109,164 $yd^3$

44. The upper barrel section of a hydrant is often called the __________ section or head.

a. nozzle
b. barrel
c. main valve
d. operating

45. If there are small rocks in the soil or the soil is unstable, it is often specified before laying down new pipe that special granular bedding material be laid down first. How thick should this granular material be?

    a. 2–4 in.
    b. 4–6 in.
    c. 8–10 in.
    d. 10–12 in.

46. These are normally located in the safety switch just ahead of the starter.

    a. Frequency relays
    b. Fuses or circuit breakers
    c. Reverse-current relays
    d. Phase reversal relays

47. Small drilling machines are used to connect service lines __________ or smaller.

    a. 2 in.
    b. 2.5 in.
    c. 4 in.
    d. 6 in.

48. The primary factor affecting power requirements for pumps is _________.

    a. pipe friction losses
    b. head
    c. amperage rating
    d. flow

49. Service lines that are more than 2 in. in diameter are usually made of _________.

    a. PVC
    b. galvanized iron
    c. high-density polyethylene
    d. the same material as water main materials

50. Compared to PVC, polyethylene pipe is _________.

    a. stronger
    b. susceptible to UV radiation like PVC
    c. expansive under the action of freezing but will not fracture
    d. not very ductile and flexible

51. What type of soil is the worst for transmitting noise?

    a. Loamy soil
    b. Clay soils
    c. Silty soils
    d. Sandy soils

52. What determines the size of a service pipe?

    a. The quantity of water required
    b. The residual pressure required by the customer
    c. The pressure in the main
    d. All of the above

53. When confirming that a well will produce at its design capacity, drawdown measurements are initially measured and recorded every __________ during the first hour, then at gradually longer intervals, typically at __________ intervals as the test progresses.

   a. 1 min, 10-min
   b. 1 min, 15-min
   c. 5 min, 15-min
   d. 5 min, 30-min

54. The bottom of a trench for drinking water installation should have _________.

   a. blocks for the pipe bells with compacted topsoil material underneath
   b. earth pads for the pipe to rest on
   c. compacted soil so the barrel of the pipe has continuous firm support
   d. clay installed under the pipe bells

55. A well produces 121 gpm. If the drawdown for the well is 13.8 ft, what is the specific yield in gpm/ft?

   a. 1,670 gpm/ft
   b. 107 gpm/ft
   c. 1.17 gpm/ft
   d. 8.77 gpm/ft

56. Butterfly valves can be used for throttling under __________ and __________ conditions.

   a. low-flow, low-pressure
   b. low-flow, high-pressure
   c. high-flow, low-pressure
   d. high-flow, high-pressure

57. In general, the rate of chlorine application for disinfecting a newly installed pipe should result in a uniform free chlorine concentration of __________ at the end of the section being treated.

   a. 2 mg/L
   b. 10 mg/L
   c. 25 mg/L
   d. 50 mg/L

58. Mechanical joints _________.

   a. do not allow very much deflection
   b. are more expensive than push-on joints
   c. take longer to assemble than push-on joints
   d. Both b and c

59. This type of pipe is no longer installed in the United States.

   a. Ductile-iron pipe
   b. Noncylinder reinforced concrete pressure pipe
   c. Gray cast-iron pipe
   d. High-density polyethylene pipe

60. In a pressure tap for a main, the cut is referred to as a __________, and it should be __________.
    a. coupon, saved for checking its condition
    b. plug, saved for checking its condition
    c. slug, discarded to an appropriate landfill
    d. slug, analyzed for chemical contamination

61. Simply flooding the backfilled or partially backfilled trench will result in good compaction only when _________.
    a. granular material with few fines is being used
    b. the native soil is being used, except if it is mostly clay
    c. a mixture of clay and silt is being used
    d. a mixture of clay and sand is being used

62. Where should vibration readings be taken?
    a. As low on the pump shaft as possible
    b. Top or bottom of pump's casing
    c. As high on the pump shaft as possible
    d. On the shaft, in or near the bearings

63. What type of hydrants have no main valve?
    a. Dry-barrel hydrants
    b. Slide-gate hydrants
    c. Wet-barrel hydrants
    d. Warm-climate hydrants

64. PVC can use bends, tees, and other fittings for __________ because they have the same outside diameter.
    a. cast-iron pipe
    b. ductile-iron pipe
    c. prestressed steel cylinder
    d. reinforced concrete pressure pipe

65. This chemical feed method of dechlorination has the best reproducible results.
    a. Gravity feed
    b. Chemical metering pump
    c. Venturi injector system
    d. Spray feed systems

66. The easiest leak to detect with an electronic amplifier is _________.
    a. a small leak
    b. a moderate leak
    c. a large leak
    d. leaks of any size using this device

67. Ground-probing radar has a disadvantage of _________.
    a. working only under certain soil conditions
    b. requiring special expertise to interpret the results
    c. being very expensive
    d. All of the above

68. At what percentage slope inclination do contractors find it easier to lay the pipe during an installation with the bells facing uphill?

   a. 6%
   b. 10%
   c. 12%
   d. 15%

69. It may take 30 min or more to fully open or close this type of large valve.

   a. Horizontal gate valve
   b. Rising stem gate valve
   c. Nonrising-stem gate valve
   d. Slide gate valve

70. Flushing a pipe after installation requires a flow velocity of at least __________ and preferably __________.

   a. 1.5 ft/s, 2.0 ft/s
   b. 2.0 ft/s, 2.5 ft/s
   c. 2.0 ft/s, 3.0 ft/s
   d. 2.5 ft/s, 3.5 ft/s

71. What type of valve is a specially designed gate valve used to connect a new water main to an existing main under pressure?

   a. Cutting-in valve
   b. Tapping valve
   c. Inserting valve
   d. Slide-in valve

72. What pressure is recommended for testing a hydrant?

   a. 150 psi
   b. 200 psi
   c. 250 psi
   d. 300 psi

73. In a pipe installation that is not yet complete, the open end should be _________.

   a. wrapped in paper to keep out dirt, groundwater, if any, and animals
   b. wrapped in plastic with plywood over the end for added strength
   c. plugged with the standard type of pipe material that is being used
   d. Screened off using a $\frac{1}{4}$-in. screen

74. What type of valve is used only when service is being initiated or discontinued?

   a. Auxiliary valves
   b. Gate valves
   c. Curb stops
   d. Corporation stops

75. Currently, what type(s) of plastic is (are) generally used for water services?

   a. Polybutylene
   b. Polyvinyl chloride
   c. Polyethylene
   d. Both b and c

76. The slug method is usually used for _________.

    a. small-diameter mains
    b. large-diameter mains
    c. PVC pipe
    d. cement or ductile-iron pipe

77. Underwater inspection of storage facilities by divers or submersible equipment should follow _________.

    a. ANSI/NSF Standard C652
    b. ANSI/NSF Standard C653
    c. ANSI/AWWA Standard C652
    d. ANSI/AWWA Standard C653

78. Before installing the sleeve around the break of a main pipe, _________.

    a. take a picture of the leak
    b. thoroughly clean the exterior of the pipe
    c. disinfect the sleeve
    d. loosen the bolts on the sleeve

79. __________ should be coated and provided with cathodic protection under __________ soil conditions.

    a. Cast iron, essentially all
    b. Steel pipe, essentially all
    c. Cast iron, most
    d. Steel pipe, most

80. __________ are provided on starters to prevent the pump motor from burning out if abnormal operating conditions increase the load beyond the pump's design.

    a. Overcurrent relays
    b. Thermal-overload relays
    c. Circuit breakers
    d. Fuses

81. According to best practices, hydrants should be set back from the curb at least _________.

    a. 2 ft
    b. 2.5 ft
    c. 3 ft
    d. 4 ft

82. The type of auxiliary valve most often used is directly connected to the hydrant by __________; one __________ of this arrangement is that the valve __________ separate from the hydrant.

    a. a flanged connection, disadvantage, can
    b. a flanged connection, advantage, cannot
    c. shackle rods, disadvantage, can
    d. shackle rods, advantage, cannot

83. Lead (Pb) service lines are now being eliminated _________.

a. whenever reasonably possible
b. after they start leaking
c. by USEPA mandate, Title 22, Section 41
d. by the USEPA Lead and Copper Rule

84. On a new pipe installation leak test, what is the preferred method for measuring makeup water volume?

a. A centrifugal pump
b. A calibrated makeup reservoir
c. A water meter
d. A pitot tube

85. During an excavation to install a water line, what do operators need to be very careful of because it often looks like a tree root?

a. Gas line
b. Galvanized pipe
c. Fiber optics
d. Electrical line

86. In regards to trench excavations, tills with low moisture content would be most similar to _________.

a. gravel
b. silts
c. clays
d. sands

87. Flare fittings are available for connecting lengths of __________ and adapting to __________. The same fittings can also be used for _________ of the same dimensions.

a. plastic pipe, iron pipe thread, ductile-iron threads
b. copper tubing, iron pipe thread, plastic pipe
c. copper tubing, ductile iron, stainless steel
d. plastic pipe, stainless steel, iron pipe

88. This type of valve is not designed to regulate or throttle water flow.

a. Globe valve
b. Gate valve
c. Plug valve
d. Needle valve

89. After about __________ of initial full-service operation on a new pump, or as directed by the manufacturer, all bearings should be regreased.

a. 2 weeks
b. 1 month
c. 2 months
d. 3 months

90. The line on the spigot end of push-on joints needs to be pushed "home" such that the line __________, when the joint is set completely, and the __________ end of the push-on joint is the end that is beveled.

    a. just disappears into the bell, spigot
    b. is all the way to the face of the bell, spigot
    c. just disappears into the bell, inside the bell
    d. is all the way to the face of the bell, inside the bell

91. After a new pipeline has passed the bacteriological tests, the highly chlorinated water should be flushed until the water exiting the pipe has a chlorine residual of _________.

    a. no greater residual than the feedwater
    b. less than 2.5 mg/L
    c. less than 4.0 mg/L
    d. less than 5.0 mg/L

92. An operator knows the water table is falling if the _________.

    a. drawdown remains the same but the static level drops
    b. drawdown is decreasing but the recovery time is increasing
    c. drawdown is increasing but the specific capacity is decreasing
    d. well yield is increasing and the recovery time is decreasing

93. The __________ thread has a __________ diameter and a steeper taper, which gives it greater strength than the __________ thread.

    a. iron-pipe, smaller, Mueller
    b. Mueller, smaller, iron-pipe
    c. Mueller, larger, iron-pipe
    d. iron-pipe, larger, Mueller

94. Critical pump equipment that is operated continuously may require _________ inspections; less critical pump equipment may be inspected _________.

    a. weekly, quarterly
    b. monthly, annually
    c. annually, biannually
    d. quarterly, annually

95. In a major leak, the first consideration is to _________.

    a. shut the line down
    b. partially or totally shut the line down
    c. try to first repair the leak with the pipe under pressure to prevent any backflow into the system
    d. inform the fire department and police

96. If a dead-end main is at least __________ in diameter and has sufficient __________, a fire hydrant should be installed __________ of the pipe.

    a. 4 in., pressure, at the end
    b. 6 in., pressure and flow, at the end
    c. 4 in., pressure and flow, near or at the end
    d. 6 in., pressure, near or at the end

97. If a pipe disinfection test fails to meet minimum standards during the first test, what should be done?

   a. The water should be tested again.
   b. Disinfect the pipe again.
   c. Add more disinfectant and test again.
   d. Repeat the entire disinfection procedure.

98. Internal combustion coolant should be checked with a __________, and if the coolant is low, __________ should be added, but never add __________.

   a. reflectometer, antifreeze, acidic water
   b. hydrometer, antifreeze and/or water, acidic water
   c. refractometer, water, acidic water
   d. hydrometer, antifreeze and/or water, alkaline water

99. What pipe quality provides for long life, toughness, and the ability to maintain tight joints with little or no maintenance?

   a. Corrosion resistance
   b. Strength
   c. Pressure rating
   d. Durability

100. After a new pipe has been laid in a trench, it should first be _________.

   a. partially backfilled
   b. completely backfilled
   c. tested to see if there are any leaks
   d. tested to see if it holds the pressure and if there are any leaks

101. The fuel level in emergency generators needs to be monitored very frequently, especially for __________ because running out of fuel can cause major damage to the __________ systems.

   a. diesel, fuel-injection
   b. gasoline, fuel pump and electrical
   c. diesel, fuel pump and electrical
   d. gasoline, fuel injection

102. Long-radius bends for steel pipe can also be made in smaller-sized pipe by the use of __________.

   a. wrinkle bending
   b. a hydraulic bender
   c. prefabricated bends by the manufacturer
   d. a hickey bender

103. Backfill for a newly installed pipe above the springline varies depending on __________.

   a. the regulatory requirements
   b. type of valves used
   c. type of connections used
   d. the utility's SOPs

104. In a trenchless main replacement, what type of installation system is used?

a. Static bursting
b. Pneumatic
c. Hydraulic
d. All of the above

105. After the first oil change for a pump motor, future oil changes should be performed every __________ depending on manufacturer recommendations, operating frequency, and __________.

a. 6–12 months, environmental conditions
b. 6–12 months, pump type
c. 1–2 years, environmental conditions
d. 1–2 years, pump type

106. To make sure there are no voids or high spots in the bedding of a trench, the operator should use __________.

a. laying blocks
b. a grading level
c. a leveling board
d. a story pole

107. When testing a hydrant at full flow, it is best to use a __________.

a. hose that is laid down in the gutter or a ditch
b. diffuser
c. rigid pipe angled down
d. gate valve

108. Well surging will dislodge fine material from the __________, and a __________ will remove the fines.

a. well face, tremie
b. well face or screen, bailer
c. screen, tremie
d. gravel pack and screen, pump

109. What is the most common cause of pipe joint failure?

a. Misalignment
b. Not clean
c. Gasket missing
d. Not fully driven "home"

110. What can an air pocket in a water main cause?

a. Cavitation
b. Water hammer
c. Restricted flow
d. Eventual corrosion in that area

111. Before pressure and leakage tests are done for a new pipe installation, allow at least __________ for the concrete used for thrust blocks to cure.

a. 2 days
b. 5 days
c. 7 days
d. 10 days

112. Storage tank screens can become clogged, usually with__________, so tanks should also have _________ to relieve excess _________ if the screens become blocked.

a. ice, flap valves, corrosion-causing chlorine moisture and pressure
b. plant debris, air-and-vacuum valves, corrosion-causing chlorine moisture and pressure
c. ice, flap valves, pressure and vacuum
d. plant debris, air-and-vacuum valves, pressure and vacuum

113. Valves stored outside in freezing climates should be __________ and have their disks in a __________ position.

a. closed, horizontal
b. open, vertical
c. closed, vertical
d. open, horizontal

114. A water audit is usually done by most utilities over a period of __________.

a. 1 week
b. 1 month
c. 1 quarter
d. 1 year

115. Electronic listening devices for finding leaks use __________.

a. geophones
b. aquaphones
c. amplifiers
d. an acoustic analyzer

116. __________ hydrants installed at the same time as a new main __________ be pressure-tested along with the main, and the hydrant's auxiliary valves should be __________ during the main pressure test.

a. Only wet-barrel, should not, closed
b. New, should not, closed
c. New, should, open
d. All, should, open

## Disinfection Monitoring, Evaluation, Adjustment, and Laboratory Analysis/Interpretation

1. A pipe that is 3.00 ft in diameter and 1,040 ft long is to be disinfected with 65.0% calcium hypochlorite [$Ca(OCl)_2$] tablets that are glued with an NSF-approved glue to the top of the pipe as it is emplaced in the trench. If the desired dose is 50.0 mg/L, how many pounds of calcium hypochlorite are required?

   a. 32.7 lb of $Ca(OCl)_2$
   b. 35.3 lb of $Ca(OCl)_2$
   c. 37.1 lb of $Ca(OCl)_2$
   d. 38.8 lb of $Ca(OCl)_2$

2. What was the primary reason that the USEPA promulgated the final Ground Water Rule in October 2006?

   a. Arsenic contamination
   b. Radon contamination
   c. Nitrites contamination
   d. Fecal contamination

3. What is the principal trihalomethane?

   a. Chloroform
   b. Bromodichloromethane
   c. Dibromochloromethane
   d. Bromoform

4. A meter being tested by a laboratory shows a reading of 118.4 $ft^3$. A volumetric tank used to measure the water that flowed through the meter indicates the actual volume as 900.1 gal. What is the percentage accuracy of the meter?

   a. 7.6% meter accuracy
   b. 13.2% meter accuracy
   c. 98.39% meter accuracy
   d. 86.8% meter accuracy

5. What water quality problem can be considered both a chemical and biological issue?

   a. Nitrification
   b. Color
   c. *E. coli*
   d. Tastes and odors

6. Nitrification can lead to disinfectant residual decay and result in growth of bacteria and a(n) _________.

   a. increase in pH
   b. increase in alkalinity
   c. reduction in oxygen
   d. reduction in nitrogen

7. Controlling lead leaching in the pH range of 6–9 pH units generally requires the dissolved inorganic carbon to be greater than _________.

   a. 2 mg/L
   b. 2.5 mg/L
   c. 3.5 mg/L
   d. 5 mg/L

8. In the Ground Water Rule, systems found to be at high risk for fecal contamination are required to provide _________ inactivation of __________.

   a. 2-log, *Cryptosporidium*
   b. 4-log, viruses
   c. 4-log, *Cryptosporidium*
   d. 2-log, viruses

9. What is the dosage in mg/L for a treatment plant that uses 318 lb/d of chlorine and treats 23.4 $ft^3/s$?

   a. 2,542.0 mg/L
   b. 0.40 mg/L
   c. 175.4 mg/L
   d. 2.52 mg/L

10. A pipe that is 2.5 ft in diameter and 4,408 ft long is to be disinfected with 12.5% sodium hypochlorite solution. If the desired dose is 25.0 mg/L, how many gallons of sodium hypochlorite are needed? Assume the hypochlorite solution weighs 10.34 lb/gal.

    a. 18.8 gal of NaOCl
    b. 23 gal of NaOCl
    c. 26 gal of NaOCl
    d. 29.6 gal of NaOCl

11. Chloramine produces a __________, and it is also, compared to chlorine, relatively __________ for __________.

    a. short-lasting residual, ineffective, inactivating most cyst-forming bacteria
    b. lower concentrations of THMs, ineffective, inactivating *Giardia*
    c. higher concentration of DBPs, ineffective, inactivating most cyst-forming bacteria
    d. long-lasting residual, effective, *Giardia*

12. How many calcium hypochlorite tablets, each weighing 0.42 lb, are needed to disinfect a water main given the following information:

| | |
|---|---|
| Length of pipe = 1,865 ft | Pipe diameter = 3.0 ft |
| Calcium hypochlorite = 65.0% available chlorine | Dosage required = 50.0 mg/L |

    a. 148 tablets of $Ca(OCl)_2$
    b. 151 tablets of $Ca(OCl)_2$
    c. 156 tablets of $Ca(OCl)_2$
    d. 163 tablets of $Ca(OCl)_2$

13. What must have been the chlorine dose if 22.6 gal of a 9.5% sodium hypochlorite solution was used to treat 725,000 gal of water?

    a. 2.96 mg/L
    b. 3.12 mg/L
    c. 3.3 mg/L
    d. 3.50 mg/L

14. What is the National Secondary Drinking Water Regulation for aluminum?

    a. 0.01–0.05 mg/L
    b. 0.05–0.2 mg/L
    c. 0.1–0.3 mg/L
    d. 0.1–0.5 mg/L

15. As a new 24.0-in.-diameter pipe that is 1,248 ft long was installed, 64.5% calcium hypochlorite [$Ca(OCl)_2$] tablets were taped to the top of the pipe. If a total of 12.25 lb were used, what must have been the dosage as calcium hypochlorite purity?

    a. 50.1 mg/L of $Ca(OCl)_2$
    b. 50.8 mg/L of $Ca(OCl)_2$
    c. 51.7 mg/L of $Ca(OCl)_2$
    d. 54.9 mg/L of $Ca(OCl)_2$

16. How many calcium hypochlorite tablets are needed to disinfect a water main given the following information:

    Length of pipe A = 1,015 ft
    Pipe A diameter = 2.00 ft
    Length of pipe B = 1,347 ft
    Pipe B diameter = 1.50 ft
    Calcium hypochlorite = 65.0% available chlorine
    Dosage required = 50.0 mg/L
    Each tablet is 0.42 lb

    a. 58 tablets
    b. 60 tablets
    c. 62 tablets
    d. 64 tablets

17. What is (are) the nitrification indicator parameter(s) that should be monitored for utilities that use chloramine?

    a. Nitrate
    b. Gram-positive coliforms
    c. Heterotrophic plate counts
    d. Ammonia-to-TOC ratio

18. The main objective of the Lead and Copper Rule is to ________.

    a. apply enhanced coagulation if out of compliance
    b. control corrosiveness of the finished water
    c. make facilities increase source water treatment for preventative measures
    d. apply cathodic protection throughout the distribution system

19. A storage tank that is going to be put back in service requires disinfection at a dosage of 50 mg/L. If the tank has a diameter of 105 ft and is 32.5 ft in height at the overflow, how many gallons of a 12.5% sodium hypochlorite (NaOCl ) solution are needed if the tank will only be filled to 10%? Give answer to the nearest gallon.

    a. 80 gal of NaOCl
    b. 81 gal of NaOCl
    c. 83 gal of NaOCl
    d. 84 gal of NaOCl

20. What parameter would not indicate a water quality problem with direct potential to impact public health?

    a. Disinfectant decay
    b. Sediment deposition
    c. Nitrification
    d. Tastes and odors

21. The frequency of sampling for contaminants, including microbiological contaminants, depends on several factors. What is one of these factors?

    a. Size of the water system
    b. Type of treatment chemicals used
    c. Treatment process
    d. Service connections

22. A disinfection profile must be prepared by water systems with TTHM or HAA5 that have annual distribution levels of __________ or __________, respectively, or higher, and they will also consist of daily __________ log inactivation.

    a. 0.60 mg/L, 0.40 mg/L, *Giardia lamblia*
    b. 0.60 mg/L, 0.40 mg/L, *Cryptosporidium*
    c. 0.64 mg/L, 0.48 mg/L, *Giardia lamblia*
    d. 0.64 mg/L, 0.48 mg/L, *Cryptosporidium*

23. This measurement (These measurements) can provide a sensitive measure of water quality in the distribution system.

    a. Chlorine residual
    b. pH
    c. Turbidity
    d. All of the above

24. Lead is hazardous if consumed in __________, particularly for __________.

    a. even very low amounts, children
    b. high amounts, children
    c. even very low amounts, adults
    d. high amounts, adults

25. A sodium hypochlorite solution is being pumped from a small day tank that is 3.5 ft in diameter. If the level in the tank drops 0.35 ft in 24.0 HR, how many milliliters per minute of hypochlorite solution was used? Make sure to use the appropriate number of significant figures in your answer.

    a. 66 mL/min
    b. 66.17 mL/min
    c. 66.2 mL/min
    d. 70 mL/min

26. What should the setting be on a chlorinator in pounds per day if the dosage desired is 3.25 mg/L and the pumping rate from the well is 735 gpm?

    a. 20.1 lb/d of chlorine
    b. 8.83 lb/d of chlorine
    c. 239.4 lb/d of chlorine
    d. 28.7 lb/d of chlorine

27. When disinfecting storage facilities using the surface application method, how long should the applied disinfection be in contact with all surfaces before potable water is introduced into the tank?

    a. 30 min
    b. 1 hr
    c. 2 hr
    d. 8 hr

28. Chloramine residuals can undergo __________ over time, releasing __________, which can promote nitrification.

    a. auto decomposition, nitric oxide
    b. biological decomposition, nitrogen
    c. auto decomposition, ammonia
    d. biological decomposition, nitrate

29. What would not require sample collection and analysis in the distribution system?

    a. Taste and odor
    b. Disinfectant residual
    c. Discoloration
    d. Water temperature approaching ambient temperature

30. The most serious example of water changing in the distribution system from the water that entered the system at the water plant is _________.

    a. bacterial growth
    b. algae growth
    c. cross-connection
    d. corrosion by products

31. With regard to the Revised Total Coliform Rule under the treatment technique for __________, __________ serves as an indicator of the potential pathway of contamination into the distribution system.

    a. coliform, total coliform
    b. fecal coliform, coliform
    c. fecal coliform, total coliform
    d. *E. coli*, coliform

32. When disinfecting a new pipeline using the chlorine tablet or granule methods, it cannot be used on _________ pipe.

    a. CIP
    b. screwed-joint steel
    c. solvent-welded plastic
    d. Both b and c

33. How many pounds of high test hypochlorite (HTH) are needed to make exactly 500 gal of a 10.0% HTH solution?

    a. 388 lb of HTH
    b. 395 lb of HTH
    c. 406 lb of HTH
    d. 417 lb of HTH

34. The Total Coliform Rule requires _________ sampling at each distribution sampling point.

    a. weekly
    b. monthly
    c. quarterly
    d. routinely, as per public health department,

35. Most violations of water regulations are caused by _________.

    a. hiding results or falsification of records
    b. not understanding requirements
    c. turning off the flow to the combined filter turbidity meter
    d. bumping the filters

36. A small tank containing 8,500 gal of water needs to be disinfected to be put back in service. If the dosage needed is 50 mg/L, how many pounds of sodium hypochlorite (NaOCl) that has 12.5% available chlorine are required?

    a. 12 lb of sodium hypochlorite
    b. 24 lb of sodium hypochlorite
    c. 28 lb of sodium hypochlorite
    d. 2,800 lb of sodium hypochlorite

37. In __________ systems, __________ formation is limited.

    a. chloramination, HAA5
    b. chloramination, TTHM
    c. chlorine dioxide, HAA5
    d. chlorine dioxide, TTHM

38. In the full storage facility chlorination method using sodium hypochlorite, the hypochlorite should be poured into the tank when the water reaches a level of _________ deep, if possible.

    a. 1–3 ft
    b. 4–5 ft
    c. 5–6 ft
    d. about 10 ft

39. If 18 lb of soda ash are mixed with 75 gal of water, what will be the percentage of soda ash in the resulting slurry?

    a. 2.8% soda ash slurry
    b. 23.35% soda ash slurry
    c. 2.9% soda ash slurry
    d. 0.28% soda ash slurry

40. What is the most important measurement of water quality in the distribution system?

    a. Turbidity
    b. Temperature
    c. pH
    d. Chlorine residual

41. What is the smallest particle that still retains the characteristics of an element?

    a. Proton
    b. Molecule
    c. Atom
    d. Nucleus

42. Forty-five pounds of soda ash is dissolved in 155 gal of water to produce a desired concentration. How many pounds of soda ash are needed to make the same desired concentration if exactly 100 gal of water is used?

    a. 29.0 lb of soda ash
    b. 69.75 lb of soda ash
    c. 14.5 lb of soda ash
    d. 3.4 lb of soda ash

43. A level __________ sanitary assessment is triggered if a public water source has __________ of the routine and repeat monthly samples as total coliform positive.

    a. 1, 5% or more
    b. 2, 5% or more
    c. 2, more than 5%
    d. 1, more than 5%

## Security, Safety, Administrative Procedures, and Public Interactions

1. If __________ tie-rods are used for restraining pipe, they should be coated or covered for protection against corrosion.

    a. stainless steel
    b. brass
    c. nonalloyed carbon steel
    d. aluminum

2. Who is required to correct workplace hazards?

    a. OSHA
    b. The employee
    c. The supervisor
    d. The employer

3. Thrust anchors are usually used when _________.

   a. soil has been disturbed such that a concrete block will not work
   b. time and cost need to be reduced
   c. there is a horizontal bend in the pipe
   d. there is a vertical bend in the pipe

4. What is the primary symptom(s) of salmonellosis?

   a. Vomiting
   b. Severe diarrhea
   c. Severe intestinal cramps
   d. High fever and diarrhea causing dehydration

5. The main reason to test hydrants and record all information is to ________.

   a. protect the public
   b. maintain the hydrants in good working order
   c. protect the operator's job
   d. Both a and b

6. When a distribution operator is confronted with the media due to a problem, it is best to _________.

   a. offer as much information as possible
   b. be detailed
   c. refer to local utility policy
   d. All of the above

7. Compliance with AWWA Standards is __________, and products __________ AWWA approved.

   a. mandatory, have never been
   b. encouraged, have never been
   c. encouraged, are always tested first to be
   d. mandatory, are always tested first to be

8. The aesthetic qualities of water include _________.

   a. taste, color, and odor
   b. DPBs and iron
   c. turbidity and pathogenic organisms
   d. pathogenic organisms, taste, and odor

9. What is the most likely threat to a drinking water system?

   a. Physical threat
   b. Biological threat
   c. Chemical threat
   d. Natural threat

10. Records indicate that the greatest number of accidents in the water industry involve _________.

    a. the weather
    b. vehicles
    c. chlorine gas
    d. confined space

11. It is of greatest importance that chemicals used to treat water do not cause _________.
    a. side effects that endanger public health
    b. tastes and odors
    c. color
    d. corrosion
12. What USEPA regulation requires consumer confidence reports to be provided to its customers?
    a. Public Notification Rule
    b. Contaminant Monitoring Rule
    c. SDWA, 1996 Amendment
    d. Surface water Treatment Rule, 1975
13. A list of vulnerabilities was determined by a __________-mandated assessment of SCADA systems for a number of large utilities in the United States.
    a. Homeland Security
    b. USEPA
    c. NSA
    d. CIA/FBI
14. Threats to a water distribution system have varying probabilities of occurring. What threat would *least* likely occur from the following?
    a. Terrorism
    b. Vandalism
    c. System aging
    d. Natural disasters
15. During an extended water outage, a temporary waterline may be used. What pipeline materials are approved?
    a. Fire hoses or rubber hoses
    b. Steel or fire hose
    c. PVC or polyethylene
    d. Polyethylene or rubber hoses
16. When traffic control is required for a pipe repair or installation, the cone taper length depends on _________.
    a. lane width
    b. traffic volume
    c. weather conditions
    d. length of time of lane closure
17. For electrical fires, use __________, but never use __________.
    a. carbon dioxide or dry powder, water or soda-acid
    b. dry powder or soda-acid, water or carbon dioxide
    c. dry powder, carbon dioxide, or soda-acid; water
    d. carbon dioxide, water or dry powder

18. Another name for shielding is __________.

    a. drag shield
    b. drag box
    c. sand shield
    d. All of the above

19. As required under the __________ Act, a drinking water utility serving more than __________ persons must do a vulnerability assessment update or complete an emergency response plan and certify these to the USEPA.

    a. Bioterrorism, 3,300
    b. Cyber Terrorism, 3,300
    c. Physical Terrorism, 10,000
    d. Chemical Terrorism, 10,000

20. The primacy agency in each state is designated by the __________, and the primacy agency can be __________.

    a. governor; the particular state's EPA
    b. USEPA; Department of Natural Resources
    c. state health department; Department of Natural Resources
    d. USEPA; the particular state's public health department

21. Wide trenches should be avoided for small-diameter pipe if at all possible, particularly in __________ soils.

    a. hard clay
    b. silty
    c. sandy
    d. hard pan

22. What regulation requires sanitary surveys?

    a. SWTR
    b. IESWTR
    c. Public Notification Rule
    d. Lead and Copper Rule

23. What soils can be easily and safely excavated?

    a. Calcareous sand
    b. Firm clays and tills with low moisture content
    c. Wet silts
    d. Sand with high moisture content

24. When digging in firm soil, there is a tendency to make the trench too __________.

    a. steep
    b. wide
    c. deep
    d. long

25. Before a confined space is entered, an employer is required to have __________.
    a. a first aid kit, defibrillator, water, and food
    b. testing, monitoring, and ventilation equipment
    c. lighting, barriers, ladders, and personal protective equipment as needed for the particular confined space
    d. Both b and c

26. Natural disasters, such as earthquakes and floods, need contingency plans that are developed by the __________.
    a. USEPA
    b. state governments
    c. utilities
    d. FEMA

27. The American Water Works Association will __________.
    a. test products
    b. develop standards
    c. approve products
    d. approve regulations

28. The major performance categories for drinking water distribution system operations are __________.
    a. to meet firefighting requirements, satisfy customer demands, and maintain water quality
    b. to efficiently perform system operations
    c. to deliver sufficient water
    d. All of the above

29. Each major item in a vulnerability assessment has critical security-related physical components. These include __________.
    a. access control
    b. power
    c. police
    d. firefighting

30. What provides the most reliable, flexible method of controlling access to a facility?
    a. 24/7 manned, closed-circuit TV
    b. Card-reader systems
    c. Key-lock systems
    d. Electrified locking systems

31. For a water utility to improve its energy use, it should have at least __________ of energy and cost data to establish a baseline.
    a. 12 months
    b. 2 years
    c. 3 years
    d. 5 years

# IV. Water Distribution, Advanced Questions

## Distribution System Components

1. The fire insurance underwriters recommend that residential areas should have minimum pipe sizes of _________.

   a. 3–4 in.
   b. 4–6 in.
   c. 6–8 in.
   d. 8–10 in.

2. If a hydrant needs to be throttled, one of the nozzles should be fitted with a __________ and the hydrant valve fully opened.

   a. plug valve
   b. ball valve
   c. butterfly valve
   d. gate valve

3. This type of water meter is used where daily flow is usually low, but, on occasion like in an emergency, the flow has to be very high.

   a. Turbo meter
   b. Acoustic meter
   c. Current meter
   d. Detector-check meter

4. An operator is filling a tank that is predominantly below ground with water. The tank is 45.0 ft in diameter and is 19.5 ft to the overflow. Estimate the time it will take to fill the tank in hours, given the following data:

   Pipe diameter = 2.00 in. Height of pipe above tank = 36.0 in.

   Distance water shoots out of pipe to top of tank (length) = 31.2 in.

   a. 65.6 hr
   b. 67.2 hr
   c. 68.7 hr
   d. 70.2 hr

5. In the pump characteristic curves, three curves are used that relate __________, __________, and __________.

   a. discharge to head, efficiency to head, power to head
   b. discharge to power, efficiency to power, head to power
   c. head to discharge, efficiency to discharge, power to discharge
   d. head to efficiency, discharge to efficiency, power to efficiency

6. The distribution system design should include equal emphasis on __________ and __________.

   a. pressure, flow velocity
   b. fire protection, domestic peak demand
   c. quantity, quality
   d. disinfection residual, disinfectant by product potential

7. What does system metering accomplish?

   a. Helps in administrating water rights
   b. Helps determine pump and pipeline capacity
   c. Helps engineering design of water system improvements
   d. All of the above

8. What type of pipe has high tensile strength, is low cost, and is good hydraulically when lined but is subject to general corrosion if installed unprotected in a corrosive environment?

   a. Steel pipe
   b. Ductile-iron pipe
   c. Cast-iron pipe
   d. Asbestos-cement pipe

9. All the power of the electric current driving the motor is used to drive the pump. However, some of the current is used to overcome _________ within the motor, and some current is lost to the conversion of _________ to __________.

   a. friction, electrical energy, mechanical energy
   b. friction, electrical energy, heat
   c. electrical resistance, potential energy, kinetic energy
   d. electrical resistance, kinetic energy, heat

10. What should be inspected on an altitude valve that fills an elevated tank?

    a. Opening and closing speed to prevent water hammer
    b. Make sure no oil leaks that could contaminate the water
    c. Make sure valve is letting enough water in to fill the tank
    d. Both a and c

11. What type(s) of large water storage is (are) usually equipped with an adjacent pumping system that can be used in an emergency to pump water to the distribution system?

    a. Elevated tanks
    b. Distribution reservoir
    c. Standpipes
    d. Pneumatic and underground tanks

12. What type of motor has a high starting torque and a high starting current?

   a. Split-phase motor
   b. Repulsion-induction motor
   c. Capacitor-start motor
   d. Synchronous motor

13. Water utilities that purchase their water during off-peak hours will usually pay much less. The rate charge for this water is often called _________.

   a. off-peak charge
   b. night charge
   c. wholesale direct charge
   d. dump charge

14. The __________ can get soil worked up from the bottom, and the __________ can probably bring the curb stop and __________ up with it if the box is pulled from the ground, for example, by a backhoe.

   a. Minneapolis-style curb stops and boxes, arch-pattern box base, meter
   b. Minneapolis-style curb stops and boxes, arch-pattern box base, service line
   c. arch-pattern base, Minneapolis-style curb stops and boxes, service line
   d. arch-pattern base, Minneapolis-style curb stops and boxes, meter

15. Calculate the total kilowatts needed to operate the following small facility if everything were running:

| | |
|---|---|
| Raw water pump | 300 hp |
| Five flocculators, 10 hp each | 60 hp |
| Filter pump for backwashing | 50 hp |
| Chlorination | 25 hp |
| Clear-water pump | 200 hp |
| Lighting | 18 hp |
| Instrumentation | 5 hp |

   a. 882 kW
   b. 487 kW
   c. 267 kW
   d. 491 kW

16. Large pumps should be started with the discharge valve _________. If a foot valve does not exist and no head is on the suction side of the pump, it should be primed using a __________.

   a. opened, vacuum pump
   b. closed, discharge line
   c. opened, discharge line
   d. closed, vacuum pump or ejector operated with steam, air, or water

17. If large capacity and close accuracy at low flows are important, what type of meter is best?

   a. Detector check-type meter
   b. Compound-type meter
   c. Proportional-type meter
   d. Positive displacement-type meter

18. If a pump operates under a high suction head, what should be used?
    a. A packing gland
    b. Lantern rings
    c. A mechanical seal
    d. Lantern rings and shaft sleeves
19. What type of concrete storage tank is constructed by first installing an inner concrete core wall that establishes the reservoir's form?
    a. Cast-in-place concrete tank
    b. Circular, prestressed concrete tank
    c. Hydraulically applied concrete-lined tank
    d. Bar-wrapped concrete tank
20. What type of actuators need sufficient reserve capacity to provide continual operation after a power failure?
    a. Power actuators
    b. Pneumatic actuators
    c. Hydraulic actuators
    d. Solenoid actuators
21. General factors to be considered for selection of pipe include service connections. What is the main factor that should be considered for service connections?
    a. Pressure
    b. Ductility
    c. Strength
    d. Cost
22. What type of transmission allows signals to pass in only one direction?
    a. Uniplex
    b. Simplex
    c. Full simplex
    d. Half simplex
23. What type of actuator will operate by a limit switch for on and off and can be fitted with controls that can be adjusted to leave the valve partially open for use in throttling?
    a. Hydraulic actuators
    b. Electric actuators
    c. Power actuators
    d. Pneumatic actuators
24. Where are air relief valves used?
    a. Next to check valves to eliminate or reduce surges
    b. On the discharge side of a well pump
    c. At the high points in transmission pipelines
    d. Both b and c

25. __________ demand can account for as much as __________ of the total capacity of a storage system.

a. Domestic, 70%
b. Fire, 50%
c. Domestic, 50%
d. Fire, 70%

26. What type of meter should have a strainer installed ahead of the flow into it to protect it from sediment and other objects in the water?

a. Magnetic meter
b. Venturi meter
c. Compound meter
d. Proportional meter

27. If sodium hypochlorite is used to disinfect a storage tank, the total volume of the entire tank should have a free chlorine residual of at least _________ after the proper detention time.

a. 5 mg/L
b. 10 mg/L
c. 25 mg/L
d. 50 mg/L

28. If a pressure test in a particular location indicates a wide variation from night to peak day use, then the ________.

a. water is being stolen, most probably from a fire hydrant
b. system is being stressed
c. system is functioning as designed
d. pumps or pressure-reducing stations are not working properly

29. A high-capacity pump will need to overcome __________ head that may be many times greater than the __________.

a. friction, pressure head
b. friction, increase in flow rate
c. velocity, pressure head
d. velocity, increase in flow rate

30. A distribution system performance category would include ________.

a. reliability
b. storage utilization
c. energy usage
d. operational procedures

31. Determine the horsepower required for a clearwell water pump that needs to pump water to a storage tank given the following parameters:

Elevation of clearwell water pump = 297.45 ft
Elevation of water storage tank = 605.11 ft
Length of pipeline from clearwell water pump to storage tank = 3,108 ft
Pump above clearwell (suction lift) = 4.52 ft
Friction loss in pipeline = 1.14 ft per 1,000 ft

Assume velocity head = 2.48 ft
Required flow per day (maximum) = 1,250 gpm
Pump efficiency = 78.6%
Motor efficiency = 87.6%

a. 146 Hp
b. 6.8 Hp
c. 175 Hp
d. 186 Hp

32. What type of actuators would use electronic solenoid valves to operate the valve in each direction?

a. Hydraulic actuators
b. Electric actuators
c. Power actuators
d. Pneumatic actuators

33. Why is the design of the suction piping so important?

a. To prevent air or water leaks into the suction side
b. To allow sufficient flow
c. To minimize pressure losses
d. Both b and c

34. This type of valve provides low-volume flow without the need for opening the main valve.

a. Needle valve
b. Bypass valve
c. Insertion valve
d. Pressure-reducing valve

35. The close fitting impeller of this type of pump prohibits any pumping of solid sediment, such as sand, fine grit, and silt.

a. Mixed-flow pump
b. Axial-flow pump
c. Vertical turbine pump
d. Radial-flow pump

36. The coatings inside a tank must not cause __________ and must meet __________.

a. tastes, ANSI standards
b. tastes and odors, NSF International standards
c. tastes and odors, AWWA/ANSI standards
d. tastes, USEPA standards

37. How does the valve work on a toggle hydrant?

a. Valve closes horizontally
b. Valve closes vertically
c. Similar to one side of an ordinary rubber-faced gate valve
d. The valve closes with the water pressure against the seat

38. How many gallons of water are in a storage tank that is 70.0 ft in radius, has a capacity of 29.2 ft, and the 4-20 mA live signal is 14.88?

   a. 305,504 gal
   b. 2,290,000 gal
   c. 3,050,000 gal
   d. 3,130,000 gal

39. In a Venturi meter, the difference in pressure before the constriction and at the throat can be measured, and the change in pressure is ________.

   a. proportional to the square root of the velocity
   b. proportional to the square of the velocity
   c. indirectly proportional to the square root of the velocity
   d. indirectly proportional to the square of the velocity

40. What is it called when signals are allowed to be passed in both directions simultaneously?

   a. Polling
   b. Full duplex
   c. Multiplexing
   d. Pulse-duration modulation

41. What type of a joint utilizes a bolted, segmental, clamp-type, mechanical coupling with a housing that encloses a U-shaped rubber gasket?

   a. Socket joints
   b. Special push-on joints
   c. Grooved joints
   d. Restrained joints

42. The fire insurance underwriters recommend high-value districts should have minimum pipe sizes of ________.

   a. 4–6 in.
   b. 6–8 in.
   c. 8–12 in.
   d. 10–14 in.

43. What type of valve is used in the distribution system where the main is not buried very deep?

   a. Insertion valves
   b. Slide valve
   c. Horizontal gate valve
   d. Resilient-seated gate valve

44. What type of pump generally develops very high heads and has low-flow capacities?

   a. Axial-flow pump
   b. Centrifugal pump
   c. Rotary pump
   d. Radial-flow pump

45. What is the starting current on a motor called?
    a. Locked rotor current
    b. Locked stator current
    c. Maximum load current
    d. Starting load current
46. What can help a water utility avoid using large pumps?
    a. Elevated storage tanks
    b. Underground storage tanks
    c. Altitude valves
    d. Pressure-reducing valves
47. What are the two main elements that affect the efficient delivery of water?
    a. Storage and plant capacities
    b. Water losses and energy usage
    c. Distribution system and pumping capacities
    d. Climate and energy usage
48. When casing is used for pipes located under a railroad or highway, the casing pipe diameter should be __________ larger than the outside diameter of the water main bells.
    a. 2–4 in.
    b. 4–6 in.
    c. 2–8 in.
    d. 4–12 in.
49. What is wire-to-water efficiency?
    a. Motor efficiency plus pump efficiency
    b. Motor efficiency multiplied by pump efficiency
    c. Motor efficiency divided by pump efficiency
    d. Pump efficiency divided by motor efficiency
50. This type of valve should be used so that restricted flow and increased pumping costs should not occur.
    a. Pressure-reducing valve
    b. Altitude valve
    c. Pressure-relief valve
    d. Air-relief valve
51. Where extreme accuracy is required, what type of level sensor can be fitted with a special adaptation that is a level transmitter called a stage recorder?
    a. Bubbler tubes
    b. Direct electronic sensors
    c. Diaphragm elements
    d. Float mechanism
52. These analog systems can be used only for long distances.
    a. Voltage and variable frequency
    b. Current and voltage
    c. Pulse-duration modulation and variable frequency
    d. Current and variable frequency

53. Double-suction pumps will use impellers _________.

 a. only of the closed design
 b. that are closed or semi-open
 c. that are semi-open or open
 d. All of the above

54. What type of pipe is generally easy to install, easy to ship, easy to handle, and generally of low cost but susceptible to buckling under a vacuum?

 a. PVC pipe
 b. Fiberglass pipe
 c. Steel noncylinder pipe
 d. Steel pipe

55. How does the valve work in a standard compression hydrant?

 a. The valve closes horizontally.
 b. The valve closes vertically.
 c. The valve closes with the water pressure against the seat.
 d. Similar to one side of an ordinary rubber-faced gate valve

56. There must be at least two poles in a motor, but there can be more depending on how the __________ is made and how the __________ are wired and connected.

 a. stator, transistors
 b. stator, current-carrying coils
 c. rotor, transistors
 d. rotor, current-carrying coils

57. An inside ladder would not be installed in elevated tanks that are _________.

 a. saucer shaped
 b. round single pedestal
 c. in cold climates
 d. in tanks without cathodic protection

58. What type of valve produces high head loss even when fully opened?

 a. Horizontal gate valve
 b. Butterfly valve
 c. Pinch valve
 d. Globe valve

59. A velocity pump will convert the velocity and pressure of the water to __________ and __________ because of the __________.

 a. low velocity, high pressure, size and shape of the piping just past the pump
 b. low velocity, high pressure, shape of the volute or shape of the diffuser vanes
 c. high velocity, low pressure, size and shape of the piping just past the pump
 d. high velocity, low pressure, shape of the volute or shape of the diffuser vanes

60. __________ head is a measurement of the __________ distance from the hydraulic grade line to the energy grade line.

 a. Velocity, vertical
 b. Pressure, vertical
 c. Pressure, vertical and horizontal
 d. Velocity, vertical and horizontal

61. The use of __________ has decreased remarkably because of structural difficulties causing premature pipe failures.

    a. polyethylene pipe
    b. cast-iron pipe
    c. polybutylene pipe
    d. cross-linked polyethylene pipe

62. What type of pipe is easy to fabricate in special configurations, relatively easy to transport and install, is resistant to shock load, and has the ability to bend to some degree but can collapse due to partial vacuum from rapid emptying?

    a. Cast-iron pipe
    b. Steel pipe
    c. Fiberglass pipe
    d. Ductile-iron pipe

63. Determine the motor horsepower for a pump station given the following parameters:

    | | | | |
    |---|---|---|---|
    | Motor efficiency: | 88.6% | Total head (TH): | 101 ft |
    | Pump efficiency: | 79.3% | Flow: | 4.22 mgd |

    a. 100 mhp
    b. 106 mhp
    c. 110 mhp
    d. 114 mhp

64. Hydraulic modeling capabilities can compute friction head loss using (the) _________.

    a. Grahams formula
    b. Pascal's Principal formula
    c. Hazen–Williams formula
    d. Boyles Law

65. What type of control is closed-loop?

    a. Proportional control
    b. Direct control
    c. Feedback control
    d. Feed forward control

66. Mechanical seals are designed so that they can be __________ balanced and have an operating life to as much as __________.

    a. hydraulically, 20,000 hr
    b. mechanically, 20,000 hr
    c. hydraulically, 80,000 hr
    d. mechanically, 80,000 hr

67. What type of piping material is corrosion resistant and lightweight and has a low installation cost but has a low modulus of elasticity?

    a. Fiberglass pipe
    b. Cast-iron pipe
    c. Cement-lined steel cylinder pipe
    d. Polyethylene pipe

68. These analog systems can be used only for short distances.

   a. Voltage and variable frequency
   b. Current and voltage
   c. Pulse-duration modulation and variable frequency
   d. Current and variable frequency

69. A valve that should not be operated frequently because it requires too much time to open or close is a ________.

   a. gate valve
   b. globe valve
   c. butterfly valve
   d. needle valve

70. Tanks that form ice inside should have ________.

   a. nothing exposed inside that can be damaged by the ice
   b. only the exposed overflow pipe
   c. only a ladder that is welded to the side of the tank
   d. only a hanging cathodic protection

71. In large motors, both the ____________________ are electromagnets.

   a. rotor and stator
   b. rotor and windings
   c. windings and induction coil
   d. induction coil and stator

72. Many older plants have instrumentation that runs on pneumatic air with an operating range of __________, and newer plants run with electrical transmission using either current or voltage, the most common being __________.

   a. 3–15 psig, 4-20 mA direct current (DC)
   b. 5–25 psig, 4-20 mA DC
   c. 3–15 psig, 4-20 mA alternating current (AC)
   d. 5–25 psig, 4-20 mA AC

73. A double-suction pump's shaft and impeller is secured together by __________. An end-suction pump's impeller is mounted on the end of the shaft and held in place by a __________.

   a. three tapered cotter pins at 120° from each other, key
   b. a key, key nut
   c. one machined curved screw that lies flat on the shaft, key
   d. two machined curved screws that lie flat on the shaft, key nut

74. Water supplies should try to reference their comprehensive maps to the __________ for range and section lines.

   a. state coordinate system
   b. US Geological Survey
   c. Department of the Interior
   d. state highway department

75. This type of control is open-loop.

    a. Feedback control
    b. Feed forward control
    c. Proportional control
    d. Reactive control

76. Proportional meters are __________ but are __________. However, they have __________.

    a. relatively accurate, difficult to maintain, little friction loss
    b. relatively accurate, inexpensive, little friction loss
    c. expensive, easy to maintain, substantial friction loss
    d. easy to maintain, expensive, substantial friction loss

77. What development got us away from manually starting large motors that posed a considerable danger of shock and electrocution?

    a. Advanced rotors and stators
    b. Better insulation
    c. Magnetic starters
    d. Advanced winding designs

78. What is another name for a transmitter?

    a. Transducer
    b. Telemetry channel
    c. Transmission channel
    d. Telemetry signal

79. Propeller meters are primarily used for the measurement of _________.

    a. water source intakes
    b. main lines
    c. intakes for large building complexes
    d. distribution intake pumps

80. What type of valve allows for the opening and closing speeds of an actuator to be adjusted?

    a. Needle valve
    b. Pressure-relief value
    c. Vacuum relief valve
    d. Bypass valve

81. What type of control reacts to an error and adjusts to correct it?

    a. Feedback control
    b. Feed forward control
    c. Proportional control
    d. Open-loop control

82. Microbial incidents due to cross-connections are most likely caused by _________.

    a. nonpotable reclaim water
    b. mortuaries
    c. medical facilities
    d. sewage

83. Water distribution system operators will occasionally deal with __________. These __________ should be referenced to __________.

a. US Geological Survey maps, geological maps, longitude and latitude
b. plan and profile drawings, engineering drawings, section lines or property lines
c. US Interior Department drawings, engineering drawings, township and ranges
d. city planning maps, engineering maps, indexes

84. What is a pump design called if it has the pump unit mounted on the shaft of the motor that drives the pump with the motor bearings supporting the impeller?

a. Close-coupled
b. Frame-mounted
c. Horizontal-case
d. Vertical-case

85. A vertical turbine pump is a type of _________.

a. velocity pump
b. axial-flow pump
c. peristaltic pump
d. positive-displacement pump

86. Vehicle load damage to buried pipe dissipates __________ ft below the surface, and the __________ the trench, the less possibility of pipe damage from vehicle loads.

a. 2–5, narrower
b. 5–10, narrower
c. 5–10, wider
d. 2–5, wider

87. The casing for a tunnel under a railroad or highway is usually made of _________.

a. ductile iron or steel
b. ductile iron or concrete pipe
c. steel or concrete pipe
d. reinforced steel and ductile iron

88. In cold climates, any exposed risers on multicolumn tanks are generally __________ in diameter or larger to allow for freezing around the inside edge of the riser.

a. 3 ft
b. 4 ft
c. 5 ft
d. 6 ft

89. What inoculant is added to molten iron that makes it much stronger in producing ductile-iron pipe?

a. Carbon
b. Nickel
c. Magnesium
d. Molybdenum

90. The clearance between the rotating impeller wear ring and the stationary case wear ring is usually _________.

    a. 0.010–0.020 in.
    b. 0.025–0.030 in.
    c. 0.020–0.025 in.
    d. 0.050–0.100 in.

91. What is a disadvantage of using a vertical turbine pump?

    a. Constant speed and flow does not guarantee uniform flow.
    b. Complex construction
    c. Large footprint for any given capacity
    d. High repair costs

## Equipment Installation, Operation, and Maintenance

1. A new development in corporation stop valves is the use of the __________ valves instead of the __________ valves.

    a. gate, ball
    b. globe, gate
    c. plug, globe
    d. ball, plug

2. What chemical used to neutralize chlorine after disinfecting a water storage tank uses the least amount in weight with all other parameters being equal?

    a. Sulfur dioxide
    b. Sodium bisulfite
    c. Sodium sulfite
    d. Sodium thiosulfate

3. To use cathodic protection for protecting steel and iron pipe, what two sacrificial anodes could be used?

    a. Brass or aluminum
    b. Brass or zinc
    c. Zinc or magnesium
    d. Magnesium or copper

4. Before a new main is put into service, what should be completed and used as a basis for the final inspection?

    a. Engineering plans
    b. As-built plans
    c. Site plans
    d. Construction plans

5. What type of hydrants have the main rod and the operating nut not sealed from water when the main valve is open?

    a. Wet-barrel hydrants
    b. Wet-top hydrants
    c. Warm-climate hydrants
    d. Flush hydrants

6. Constant pumping rate changes would most likely occur if a utility had _________.

   a. elevated storage facilities
   b. ground-level storage facilities
   c. underground storage facilities
   d. Both a and b

7. If a newly installed pipe section needs flushing but there is insufficient water or the pipe is too large, it can be cleaned using _________.

   a. polypigs
   b. jetting with air/water mix
   c. cloth swabs
   d. foam swabs

8. On thermocouples, the temperature changes between __________ cause __________ to be generated, which can be read out directly or amplified through a __________.

   a. a point on each wire, voltage, transistor
   b. two points, amperage, transducer
   c. a point on each wire, amperage, transistor
   d. two points, voltage, transducer

9. When good-quality material is used to backfill a trench for a new pipe up to the springline, it is best to add a second layer of the same material that is at least _________ over the top of a pipe that is 8 in. in diameter and more if a larger pipe is being covered.

   a. 4 in.
   b. 6 in.
   c. 8 in.
   d. 12 in.

10. How many pounds of lime are needed to make 10.0% slurry if exactly 100 gal of slurry are required?

   a. 90.9 lb of lime
   b. 92.7 lb of lime
   c. 93.0 lb of lime
   d. 93.4 lb of lime

11. What is the curing time for a pipe that is lined with epoxy resin?

   a. At least 4 hours
   b. At least 12 hours
   c. At least 16 hours
   d. At least 24 hours

12. If a main cannot be buried deep enough to avoid frost penetration, what works well is insulating them with __________.

   a. fiberglass insulation
   b. closed-cell Styrofoam insulation board
   c. bead-board
   d. heat tape

13. Cured-in-place lining of a pipe can use a tube composed of ________.
    a. low-density polyethylene
    b. polypropylene
    c. woven polyester
    d. polystyrene

14. Selecting machines that are too large for trench excavations will most probably cause _________.
    a. side-slope instability
    b. subsequent settling over former trench
    c. unnecessary damage
    d. prolonged public inconvenience

15. When two or more elements are bonded together to form a(n) __________, the resulting particle is called a __________.
    a. covalent bond, molecule
    b. compound, molecule
    c. ionic bond, compound
    d. emulsion, mixture

16. When a __________ valve is used with a vertical pump, an air-release valve equipped with __________ can be used.
    a. surge anticipator, a vacuum breaker
    b. surge anticipator, a high-pressure sensor
    c. pump control, a vacuum breaker
    d. pump control, a high-pressure sensor

17. __________ protection systems use an external source of __________ current that makes the structuree __________ protected with respect to some other metal in the ground.
    a. Impressed-current cathodic, alternating, anodic
    b. Impressed current cathodic, direct, cathodic
    c. Galvanic, alternating, anodic
    d. Galvanic, direct, cathodic

18. What type of pump requires careful installation to ensure proper alignment of all shafting and impeller stages throughout its length and has the drive unit located at the surface?
    a. Submersible pump
    b. Deep-well pump
    c. Booster pump
    d. Centrifugal-jet pump

19. During main construction, roads can get dusty. What can usually help reduce the amount of dust on a gravel road?
    a. Salt
    b. Magnesium carbonate
    c. Calcium sulfate
    d. Calcium

20. What chemical used to neutralize chlorine after disinfecting a water storage tank uses the most amount in weight with all other parameters being equal?

    a. Sulfur dioxide
    b. Sodium bisulfite
    c. Sodium sulfite
    d. Sodium thiosulfate

21. What type of analog telemetry signaling has a protocol in which the time that a signal pulse remains on varies with the value of the parameter being measured?

    a. Feedback control loop
    b. Feed forward control
    c. Pulse-duration modulation
    d. Proportional control

22. With what soil type(s) is it best to use irregular drum tampers?

    a. Clays
    b. Clays, tills, and silts
    c. Fine-to-coarse sands
    d. Coarse sands and gravel

23. What is the most expensive part of installing distribution piping?

    a. The labor
    b. The excavation
    c. The pipe and its accessories
    d. The equipment used

24. If a new pipe installation leakage test measured on the next day is greater than the previous test day, the leak is probably in a __________. If the leakage is the same, it probably is from a __________.

    a. pipe joint or valve, service connection
    b. service connection or a damaged pipe, valve or pipe joint
    c. pipe joint or a damaged pipe, valve or service connection
    d. pipe joint or service connection, valve

25. What method of disinfecting a pipe will not work if, after the pipe is installed, it needs to be flushed?

    a. Slug method
    b. Continuous feed
    c. Tablet method
    d. Liquid chlorine method

26. A failed leakage test may be due to the saturation time being too short. What should the saturation time be for a concrete pipe?

    a. 6 hours
    b. 12 hours
    c. 18 hours
    d. 24 hours

27. Find the drawdown of a well if the well yields 206 gpm and the specific yield is 14.8 gpm/ft.

   a. 13.2 ft
   b. 13.9 ft
   c. 14.3 ft
   d. 14.7 ft

28. What type of hydrants are designed to reduce the possibility of the threads becoming fouled by sediment or erosion?

   a. Flush hydrants
   b. Dry-top barrel hydrants
   c. Warm-climate hydrants
   d. Wet-barrel hydrants

29. What type of valve is installed in a water main without shutting the pressure off but needs special equipment to make the installation?

   a. Cutting-in valve
   b. Tapping valve
   c. Inserting valve
   d. Slide valve

30. The gravel pack for a well must be clean, washed, and composed of __________ particles that are __________ times __________ than the median size of the surrounding native material.

   a. well-rounded, two to three, larger
   b. well-rounded, four to five, larger
   c. angular, two to three, smaller
   d. angular, four to five, smaller

31. If wood shoring needs to be left in place during the backfilling of a trench, it is usually cut off __________ below the surface.

   a. 0.5 ft
   b. 1 ft
   c. 2 ft
   d. 3 ft

32. The amount of current on a buried pipe is __________ to the electrical resistivity of the soil.

   a. proportional
   b. inversely proportional
   c. indirectly proportional
   d. equal

33. After a pipeline section has passed the bacteriological tests, it should be flushed to remove the high concentration of chlorine. How many pipe volumes for this particular section should be used to flush the pipe adequately?

   a. One to two pipe volumes
   b. Two to three pipe volumes
   c. Three to four pipe volumes
   d. Four to five pipe volumes

34. If a freshly cleaned pipe is to be epoxy lined, the temperature of the pipe must be above ________.

    a. 38°F (3.3°C)
    b. 40°F (4.4°C)
    c. 42°F (5.6°C)
    d. 45°F (7.2°C)

35. When operators observe vibrations or overheating of a bearing or ________, they should inspect the pump's ________ and ________ immediately.

    a. coupling, shaft coupling, packing
    b. motor windings, shaft, seals
    c. motor windings, shaft coupling, packing
    d. coupling, impeller, bearings

36. How should stringing a pipe at the job site be done?

    a. String pipe bells in the direction water will flow.
    b. If possible vandalism, string only enough for two days of work.
    c. Keep pipe ends covered to prevent contamination.
    d. Install the gaskets at least five pipe lengths ahead of the work.

37. What operational practice could adversely affect distribution system water quality performance?

    a. External corrosion control
    b. Internal corrosion control
    c. Nitrification control
    d. Sediment control

## Disinfection Monitoring, Evaluation, Adjustment, and Laboratory Analysis/ Interpretation

1. Why do the 5 haloacetic acids (HAA5) degrade over time in the distribution system?

    a. Because microorganisms biodegrade them
    b. Because the hydrogen on the carboxylic group is lost to pipe corrosion
    c. Because the carboxylic group is lost due to pH values >7.2, which occurs in most distribution pipe
    d. Because they react with the alkalinity in the distribution system

2. When using the slug method to disinfect a new pipeline, if the pH goes above ________, virtually all the chlorine residual is present as hypochlorite ion, which is much less effective than hypochlorous acid.

    a. 9.5
    b. 10
    c. 10.5
    d. 11

3. Utilities that use chloramines as __________ disinfectants should establish __________ and have response plans to control nitrification.

   a. primary, an MCL
   b. secondary, action levels
   c. primary, action levels
   d. secondary, an MCL

4. If bacterial regrowth and disinfectant by products are persistent in the distribution system, what is recommended as a possible solution?

   a. Carbon filtration
   b. Ozone biofiltration
   c. Source water treatment
   d. Enhanced coagulation

5. How many gallons of a sodium hypochlorite solution that contains 11.7% available chlorine are needed to disinfect a water main (where the main is 14.0 in. in diameter and the pipeline is 1,002 ft long) if the dosage required is 25.0 mg/L? Assume the sodium hypochlorite solution weighs 10.12 lb/gal.

   a. 1.40 gal of sodium hypochlorite
   b. 1.41 gal of sodium hypochlorite
   c. 1.44 gal of sodium hypochlorite
   d. 1.47 gal of sodium hypochlorite

6. A mole of a substance is defined as the number of grams of that substance for which the number equals the substance's _________.

   a. equivalent weight
   b. molecular weight
   c. valence number multiplied by the equivalent weight
   d. molality divided by valence number

7. Under wet-trench construction, water entering the pipe as it is installed cannot be helped in some circumstances. This water should be treated with chlorine granules or tablets so the water has a chlorine residual of _________.

   a. 10 mg/L
   b. 25 mg/L
   c. 50 mg/L
   d. 100 mg/L

8. The radionuclide in drinking water that is most responsible for the annual dose received by the average person is _________.

   a. radium 226
   b. radium 228
   c. uranium
   d. radon

9. A solution containing exactly 825 gal of 6.00% sodium hypochlorite solution is required. How many gallons of a 12.5% solution must be mixed with a 3.30% solution to make the required solution? Give answers to nearest gallon.

   a. Mix 235 gal of the 12.5% solution with 590 gal of the 3.30% solution
   b. Mix 237 gal of the 12.5% solution with 588 gal of the 3.30% solution
   c. Mix 242 gal of the 12.5% solution with 583 gal of the 3.30% solution
   d. Mix 246 gal of the 12.5% solution with 579 gal of the 3.30% solution

10. What National Secondary Drinking Water Regulation contaminant is associated with corrosion, hardness, scale, and taste?

   a. TDS
   b. $H_2S$
   c. Sulfate
   d. Foaming agents

11. A water treatment plant has a filter flow of 3,205 gpm and is being treated with 215 gpd of a hypochlorite solution weighing 10.3 lb/gal. If the desired dose is 5.2 mg/L (demand plus residual), determine the concentration of the hypochlorite solution in percentage.

   a. 9.0%
   b. 9.2%
   c. 9.3%
   d. 9.5%

12. A pipe that is 1.50 ft in diameter and 1,251 ft long is to be disinfected with 12.5% sodium hypochlorite (NaOCl) solution. If the desired dose is 25 mg/L, how many gallons of sodium hypochlorite are needed? The sodium hypochlorite solution weighs 10.32 lb/gal.

   a. 2.6 gal of NaOCl
   b. 2.7 gal of NaOCl
   c. 3.0 gal of NaOCl
   d. 3.3 gal of NaOCl

13. What is to be used to classify surface water sources into one of four USEPA-defined risk levels called "bins" in the Long-Term Surface Water Treatment Rule?

   a. Coliforms
   b. *E. coli*
   c. *Cryptosporidium*
   d. *Giardia lamblia*

14. How many gallons of a sodium hypochlorite solution that contains 11.5% available chlorine are needed to disinfect a 1.5-ft-diameter pipeline that is 412 ft long if the dosage required is 25.0 mg/L? Assume the sodium hypochlorite is 9.25 lb/gal.

   a. 1.07 gal of sodium hypochlorite
   b. 1.41 gal of sodium hypochlorite
   c. 1.2 gal of sodium hypochlorite
   d. 1.31 gal of sodium hypochlorite

15. What is the density of a polymer solution in grams per cubic centimeters if the polymer solution weighs 50.5 lb and occupies 5.00 gal?

    a. 4.58 g/cm$^3$
    b. 0.83 g/cm$^3$
    c. 1.21 g/cm$^3$
    d. 10.1 g/cm$^3$

16. What is the most important factor in determining which atoms will combine with other atoms?

    a. The electron shells
    b. The family it is in on the periodic table
    c. The valence electrons
    d. Number of protons and neutrons

17. If 7 lb, 12 oz of soda ash is added to 50.0 gal of water, what is the percentage of soda ash in the slurry?

    a. 1.78% soda ash slurry
    b. 1.80% soda ash slurry
    c. 1.2% soda ash slurry
    d. 1.86% soda ash slurry

18. How many pounds per day of chlorine gas are required to treat 5.36 mgd if the dosage is 3.25 mg/L?

    a. 145 lb/d of chlorine
    b. 150 lb/d of chlorine
    c. 152 lb/d of chlorine
    d. 153 lb/d of chlorine

19. The major source of organic chemicals in drinking water is from _________.

    a. industrial and synthetic wastes
    b. domestic waste, including landfills and urban runoff
    c. reactions that take place during water treatment and distribution
    d. breakdown of naturally occurring organic chemicals

20. In a pipe repair using the slug method, the chlorine concentration of 300 mg/L can have a contact time as short as _________.

    a. 15 min
    b. 30 min
    c. 1 hour
    d. 2 hours

21. If a new main floods during a storm, it should be flushed out and chlorinated with water containing __________, which should remain for __________.

    a. 25 mg/L, 24 hours
    b. 50 mg/L, 24 hours
    c. 25 mg/L, 48 hours
    d. 50 mg/L, 48 hours

22. When using the tablet method to disinfect a pipeline, if the temperature of the water is less than ________, the water must be held in contact with the pipe for at least ________.

    a. 38°F (3.3°C); 24 hours
    b. 41°F (5°C); 24 hours
    c. 38°F (3.3°C); 48 hours
    d. 41°F (5°C); 48 hours

23. A liquid polymer weighs 4.35 lb and occupies 1.559 L. What is the density of the polymer in grams per cubic centimeters?

    a. 1.25 g/cm$^3$
    b. 1.27 g/cm$^3$
    c. 1.3 g/cm$^3$
    d. 1.32 g/cm$^3$

24. If a pipeline has passed all the bacteriological tests and one pipe needs to be installed to make the final connection to the working distribution system, this last pipe can be disinfected with a(n) ________ chlorine solution just before installation.

    a. 1% to 5%
    b. 5% to 9%
    c. 9% to 12.5%
    d. at least 15%

25. A 2.0-MG storage tank needs to be disinfected with calcium hypochlorite, $Ca(OCl)_2$, granules that contain 63.0% available chlorine. The tank is to be filled at 10% capacity, and the initial chlorine dosage desired is 25.0 mg/L. How many pounds of calcium hypochlorite are required?

    Give the answer to two significant figures.

    a. 66 lb $Ca(OCl)_2$
    b. 66.19 lb $Ca(OCl)_2$
    c. 66.2 lb $Ca(OCl)_2$
    d. 70 lb $Ca(OCl)_2$

26. The SI unit commonly used to express and compare the biological effects of radiation from different sources is the ________, where one ________ is equal to ________ rem.

    a. sievert, sievert, 10
    b. sievert, sievert, 100
    c. roentgen, roentgen, 10
    d. roentgen, roentgen, 100

27. A well that is 263 ft deep and 10.0 in. in diameter requires disinfection. Depth to water from top of casing is 92.5 ft. If the desired dose is 50.0 mg/L, how many gallons of calcium hypochlorite (27.5% available chlorine after dissolving the granules) are required? The calcium hypochlorite is 10.42 lb/gal.

    a. 0.096-gal $Ca(OCl)_2$
    b. 0.100-gal $Ca(OCl)_2$
    c. 0.101-gal $Ca(OCl)_2$
    d. 0.103-gal $Ca(OCl)_2$

28. After a main pipeline has been repaired, bacteriological samples must be collected __________, and if a positive total coliform test is obtained, additional samples must be collected until __________ negative results are obtained.

    a. at the repair site and after the repair site, two consecutive
    b. before and after the repair site, two consecutive
    c. at the repair site and after the repair site, three consecutive
    d. before, at the repair site, and after the repair site; three consecutive

29. When disinfecting a storage facility using the full storage chlorination method, the chlorinated water should be retained for at least __________, when the continuous feed is used from a gas-chlorinator or chemical feed pump.

    a. 6 hours
    b. 12 hours
    c. 18 hours
    d. 24 hours

30. When an acid and a base are mixed, the process is called _________.

    a. Bronsted–Russell reactions
    b. decomposition reactions
    c. neutralization
    d. salt synthesis

31. What is the percentage strength of a solution if 245 gal of an 11.9% solution is mixed with 136 gal of a 2.90% solution? The 11.9% solution has a specific gravity of 1.22, and the 2.90% solution weighs 9.86 lb/gal.

    a. 8.31% strength of new solution
    b. 8.43% strength of new solution
    c. 8.61% strength of new solution
    d. 8.75% strength of new solution

32. Determine the specific gravity of a substance that is 33.08 kg/ft$^3$.

    Note: One cubic foot = 28,317 cm$^3$.

    a. 0.001
    b. 0.856
    c. 1.168
    d. 1.2

33. Some utilities have employed booster disinfection in remote areas of the distribution system. What does this cause?

    a. Increases total disinfection costs
    b. Increases disinfection by products
    c. Provides more uniform disinfection at a lower level
    d. Requires less maintenance

34. A soda ash solution was added to the effluent pipe of a distribution storage tank that is 120 ft in diameter. The soda ash was injected into the pipeline at an average rate of 41.2 g/min. If the average pumping rate into the tank and into the distribution system over the past 24 hours was 2,905 gpm, and the level in the tank increased by 4.18 ft, what was the average dosage in milligrams per liter?

    a. 4.09 mg/L of soda ash
    b. 4.21 mg/L of soda ash
    c. 4.3 mg/L of soda ash
    d. 4.31 mg/L of soda ash

35. What disinfection by product degrades over time due to bacterial action?

    a. Bromate
    b. Nitrosamines
    c. THMs
    d. HAA5

36. In the slug method of disinfecting a new pipeline, if at any point the sampling result indicates a chlorine residual of less than __________, stop the procedure and add chlorine to the slug area to restore free chlorine residual to at least __________.

    a. 50 mg/L, 100 mg/L
    b. 50 mg/L, 200 mg/L
    c. 100 mg/L, 200 mg/L
    d. 100 mg/L, 300 mg/L

## Security, Safety, Administrative Procedures, and Public Interactions

1. Which of the following is a chemical warfare agent?

    a. Arsenic fluoride
    b. Dieldren
    c. Tabun
    d. Parathion

2. What is lacking if a supervisor micromanages his or her employees?

    a. Teamwork
    b. Communication
    c. Confidence in the available talent
    d. Efficiency

3. Materials used for potable water pipes, such as sealing materials, gaskets, and lubricants, should be certified according to _________.

    a. NSF Standard 60
    b. ANSI/NSF Standard 60
    c. ANSI/NSF Standard 61
    d. NSF Standard 61

4. __________ are usually either timbers or __________.

    a. Uprights, metal plates
    b. Stringers, metal plates
    c. Trench braces, adjustable trench jacks
    d. Stringers, adjustable trench jacks

5. Who is ultimately responsible for ensuring employees are trained and advised of relevant situations?

   a. The employee
   b. The supervisor of that employee
   c. The employee's union
   d. The operational manager

6. An industrial chemical poison would be _________.

   a. sodium fluoroacetate
   b. polyphosphate
   c. 3-quinucli dinyl benzilate
   d. lysergic acid diethylamide

7. Permanent and conspicuous warning signs should be installed for panels carrying more than _________.

   a. 400 V
   b. 400 A
   c. 600 V
   d. 600 A

8. Collapsible shoring assemblies are __________ panels of __________ or ___________ that are available in various panel sizes.

   a. lightweight, steel, aluminum
   b. lightweight, wood, steel
   c. heavy, iron, wood
   d. heavy, steel, wood

9. One of the most important aspects of a supervisor's job is ________.

   a. increasing output
   b. decreasing production costs
   c. training
   d. planning

10. Temporary pipelines used during an extended water outage also need to be chlorinated according to ________.

    a. ANSI/NSF Standard C650
    b. ANSI/NSF Standard C651
    c. ANSI/AWWA Standard C650
    d. ANSI/AWWA Standard C651

# I. Water Treatment, Introductory Answers

## Treatment Process

1. **b.** 12,266 $ft^2$

   Reference: *Math for Water Treatment Operators (AWWA)*

   Equation: Area $= \pi r^2$, where $\pi = 3.14$

   First, find the radius: Radius, ft = Diameter, ft/2 = 125 ft/2 = 62.5 ft

   Area of circular pond = (3.14) (62.5 ft) (62.5 ft) = 12, 265.625 $ft^2$, round to 12,266 $ft^2$

2. **a.** 361 gpd/$ft^2$

   Reference: *Math for Water Treatment Operators (AWWA)*

   First, convert the number of $ft^3$/s to gallons per day.

   $\left(4.75\,ft^3/s\right)(86,400\,s/d)\left(7.48\,gal/ft^3\right) = 3,069,792\,gal/d$

   Equation: Surface loading rate $= \dfrac{\text{gallons per day (gpd)}}{\text{Number of ft}^2}$

   Surface loading rate $= \dfrac{3,069,792\,gpd}{(125)(68\,ft)} = 361.152\,gpd/ft^2$, round to 361 gpd/$ft^2$

3. **a.** Chloroform

   Reference: *Water System Operations (WSO) Series (AWWA)*

4. **b.** 4.6 ft/s

   Reference: *Math for Water Treatment Operators (AWWA)*

   First, convert the number of gallons per minute to cubic feet per second.

   Number of $ft^3/s = \dfrac{178\,gpm}{\left(7.48\,gal/ft^3\right)(60\,s/min)} = 0.3966\,ft^3/s$

   Next, convert the diameter from inches to feet.

   Number of ft = (4.0 in.) (1 ft/12 in.) = 0.333

   Equation: Flow, $ft^3/s = \left(\text{Area, } ft^2\right)$ (Velocity, ft/s), where the area, $ft^2$ = (0.785) (Diameter, ft)$^2$

   $0.3966\,ft^3/s = (0.785)(0.333, ft)(0.333, ft)(\text{Flow, ft/s})$

   Rearrange and solve for the flow in feet per second.

   Flow, ft/s $= \dfrac{0.3966\,ft^3/s}{(0.785)(0.333)(0.333)} =$ 4.556 ft/s, round to 4.6 ft/s

5. **c.** Blue-green algae
   Reference: *Water System Operations (WSO) Series (AWWA)*

6. **b.** Coarser sand
   Reference: *Water System Operations (WSO) Series (AWWA)*

7. **c.** 98.5 mL/min of polymer
   Reference: *Math for Water Treatment Operators (AWWA)*

   First, determine the pounds per gallon of the polymer solution.

   $$\text{Polymer, lb/gal} = (\text{Specific gravity})(8.34\ \text{lb/gal}) = (1.28)(8.34\ \text{lb/gal}) = 10.675\ \text{lb/gal}$$

   Next, find the number of pounds per day of polymer required by using the "pounds" equation.

   $$\text{Polymer, lb/d} = (\text{mgd})(\text{Dosage, mg/L})(8.34\ \text{lb/gal})$$

   $$\text{Polymer, lb/d} = (13.7\ \text{mgd})(3.50\ \text{mg/L})(8.34\ \text{lb/gal}) = 399.9\ \text{lb/d}$$

   Next, convert the number of lb/d to number of gpd.

   $$\text{Polymer, gpd} = \frac{399.9\ \text{lb/d}}{10.675\ \text{lb/gal}} = 37.46\ \text{gpd}$$

   Then convert gpd to mL/min.

   $$\text{Polymer, gpd} = \frac{(37.46\ \text{gpd})(3{,}785\ \text{mL/gal})}{1{,}440\ \text{min/d}} = 98.5\ \text{mL/min of polymer}$$

8. **a.** stabilization, a slime layer
   Reference: *Water System Operations (WSO) Series (AWWA)*

9. **c.** low, low
   Reference: *Basic Microbiology for Drinking Water (AWWA)*

10. **a.** Slow sand filters
    Reference: *Water System Operations (WSO) Series (AWWA)*

11. **a.** 4.4 gpm/ft
    Reference: *Math for Water Treatment Operators (AWWA)*

    Equation: $\text{Specific yield, gpm/ft} = \frac{\text{Well yield, gpm}}{\text{Drawdown, ft}}$

    $$\text{Specific yield, gpm/ft} = \frac{97\ \text{gpm}}{22\ \text{ft}} = 4.41\ \text{gpm/ft, round to } 4.4\ \text{gpm/ft}$$

12. **c.** Monochloramine and dichloramine
    Reference: *Water System Operations (WSO) Series (AWWA)*

13. **d.** 16 mg/L Mg hardness as $CaCO_3$
    Reference: *Math for Water Treatment Operators (AWWA)*

    Rearrange the equation to solve for magnesium hardness.

    Mg hardness, mg/L as $CaCO_3$ = Total hardness as mg/L $CaCO_3$ − Ca hardness as mg/L $CaCO_3$

    Mg hardness, mg/L as $CaCO_3$ = 97 mg/L total hardness mg/L as $CaCO_3$ − 81 mg/L Ca hardness mg/L as $CaCO_3$

    Mg hardness, mg/L as $CaCO_3$ = 16 mg/L of Mg hardness as $CaCO_3$

14. **a.** 1.0–1.2 times the iron concentration
Reference: *Water Treatment Operator Training Handbook (AWWA)*
15. **d.** THMs
Reference: *Water System Operations (WSO) Series (AWWA)*
16. **d.** All of the above
Reference: *Water System Operations (WSO) Series (AWWA)*
17. **c.** Very fine silt
Reference: *Water System Operations (WSO) Series (AWWA)*
18. **d.** chlorine
Reference: *Water System Operations (WSO) Series (AWWA)*
19. **d.** All of the above
Reference: *Water Treatment Operator Training Handbook (AWWA)*
20. **c.** 60 μ g/L
Reference: *Water System Operations (WSO) Series (AWWA)*
21. **b.** Storage bin or hopper to empty of sand and grit
Reference: *Water System Operations (WSO) Series (AWWA)*
22. **c.** Sequestering agents
Reference: *Water System Operations (WSO) Series (AWWA)*
23. **c.** bed expansion, quarterly
Reference: *Water Treatment Operator Training Handbook (AWWA)*
24. **d.** Both b and c
Reference: *Water System Operations (WSO) Series (AWWA)*
25. **a.** Largest pore opening
Reference: *Water System Operations (WSO) Series (AWWA)*
26. **c.** Backwashed with permanganate solution
Reference: *Water System Operations (WSO) Series (AWWA)*
27. **a.** 325 min
Reference: *Math for Water Treatment Operators (AWWA)*

First, convert mgd to gpm, as detention time is asked for in minutes.

$$(8.48\text{ mgd})(1,000,000/1\text{M})(1\text{ day}/1,440\text{ min}) = 5,889\text{ gpm}$$

Next, determine the volume in gallons for the clarifier.

$$\text{Volume, gal} = (0.785)(\text{Diameter, ft})^2(\text{Depth, ft})\left(7.48\text{ gal/ft}^3\right)$$

$$\text{Volume, gal} = (0.785)(150.0\text{ ft})(150.0\text{ ft})(14.5\text{ ft})\left(7.48\text{ gal/ft}^3\right) = 1,915,675\text{ gal}$$

$$\text{Equation: Detention time, min} = \frac{\text{Volume, gal}}{\text{Flowrate, gpm}}$$

$$\text{Detention time, min} = \frac{1,915,675\text{ gal}}{5,889\text{ gpm}} = 325.297\text{ min, round to }325\text{ min}$$

28. **a.** packed tower aeration
Reference: *Water System Operations (WSO) Series (AWWA)*

29. **b.** Longer filter run times

    Reference: *Water System Operations (WSO) Series (AWWA)*

30. **a.** 4.3°C

    Reference: *Math for Distribution System Operators (AWWA)*

    Equation: °C = 5/9(°F − 32°F)

    °C = 5/9(39.7°F − 32°F) = 5/9(7.7°C) = 4.3°C

31. **c.** Foreign substances

    Reference: *Water System Operations (WSO) Series (AWWA)*

32. **b.** calcium hydroxide, lime

    Reference: *Water System Operations (WSO) Series (AWWA)*

33. **a.** Detention time

    Reference: *Water System Operations (WSO) Series (AWWA)*

34. **d.** 123,000 gal

    Reference: *Math for Water Treatment Operators (AWWA)*

    First, convert kilograins to grains:

    $$\left(24.8\,\text{kilograins/ft}^3\right)(1,000\,\text{grains/kilograins}) = 24,800\,\text{grains/ft}^3$$

    Next, find the capacity of the unit in grains.

    $$\left(102\,\text{ft}^3\right)\left(24,800\,\text{grains/ft}^3\right) = 2,529,600\,\text{grains}$$

    $$\text{Equation: Water treatment capacity, gal} = \frac{(\text{Exchange capacity in grains})}{\text{Hardness, gpg}}$$

    Water treatment capacity, gal = 2,529,600 grains/20.6 gpg = 122,796 gal, round to 123,000 gal

35. **d.** All of the above

    Reference: *Water System Operations (WSO) Series (AWWA)*

36. **a.** coagulation, flocculation, and sedimentation

    Reference: *Water System Operations (WSO) Series (AWWA)*

37. **c.** 6.0

    Reference: *Water System Operations (WSO) Series (AWWA)*

38. **b.** pH

    Reference: *Water Treatment Operator Training Handbook (AWWA)*

39. **b.** Because installation and operating costs are low

    Reference: *Water System Operations (WSO) Series (AWWA)*

40. **a.** high pH, low solubility, precipitate

    Reference: *Water System Operations (WSO) Series (AWWA)*

41. **b.** 47.5 ft$^3$/s

    Reference: *Math for Water Treatment Operators (AWWA)*

    $$\frac{(30.7\,\text{mgd})(1,000,000\,\text{gal})(1\,\text{ft}^3)(1\,\text{day})(1\,\text{min})}{1\,\text{MG}\quad(7.48\,\text{gal})(1,440\,\text{min})(60\,\text{s})} = 47.5\,\text{ft}^3/\text{s}$$

42. **c.** 0.2 mg/L

Reference: *Water System Operations (WSO) Series (AWWA)*

43. **c.** ultrafiltration and multimedia granular filtration

Reference: *Water System Operations (WSO) Series (AWWA)*

44. **b.** algae

Reference: *Water System Operations (WSO) Series (AWWA)*

45. **c.** Use a coagulant aid

Reference: *Water System Operations (WSO) Series (AWWA)*

46. **c.** Soda ash

Reference: *Water System Operations (WSO) Series (AWWA)*

47. **c.** alkalinity, 0.5 mg/L

Reference: *Water System Operations (WSO) Series (AWWA)*

48. **c.** inadequate coagulant dosage

Reference: *Water System Operations (WSO) Series (AWWA)*

49. **d.** 60.0 gpm

Reference: *Math for Water Treatment Operators (AWWA)*

First, determine the number of minutes the pump was working:

$$(1\,\text{hr})\,(60\,\text{min/h}) + 52\,\text{min} = 60\,\text{min} + 52\,\text{min} = 112\,\text{min}$$

Then, determine the number of gpm by dividing the number of gallons pumped by the total time the pump worked:

$$\frac{6,720\,\text{gal}}{112\,\text{min}} = 60.0\,\text{gpm}$$

50. **c.** 718 lb/ft$^2$

Reference: *Math for Water Treatment Operators (AWWA)*

Equation using pressure in lb/ft$^2$ is Pressure, lb/ft$^2$ = (Depth, ft) (Density, 62.4 lb/ft$^3$)

$$\text{Pressure} = (11.5\,\text{ft})\left(62.4\,\text{lb/ft}^3\right) = 717.6\,\text{lb/ft}^2\text{, round to } 718\,\text{lb/ft}^2$$

51. **a.** Positive-displacement meters

Reference: *Water System Operations (WSO) Series (AWWA)*

52. **d.** 50% final concentration

Reference: *Math for Water Treatment Operators (AWWA)*

First, calculate the final volume ($V_2$) in mL.

$$V_2 = 250\,\text{mL} + 100\,\text{mL} = 350\,\text{mL}$$

$C_1V_1 = C_2V_2$ Substitute known values and solve.

$$(70\%)\,(250\,\text{mL}) = (C_2)\,(350\,\text{mL})$$

$$C_2 = \frac{(70\%)\,(250\,\text{mL})}{350\,\text{mL}} = 50\%\text{ Final concentration}$$

53. **a.** pinhead-size floc
Reference: *Water System Operations (WSO) Series (AWWA)*

54. **b.** To maintain a uniform sludge layer, 40–50 s
Reference: *Water System Operations (WSO) Series (AWWA)*

55. **c.** 8,000 gpd/ft
Reference: *Math for Water Treatment Operators (AWWA)*

First, change 2.71 mgd to gpd:(2.71 mgd)(1,000,000/1 M) = 2,710,000 gpd

Next, determine the total weir length.

Total weir length, ft = (4 clarifiers)(85 ft/each) = 340 ft

Equation: Weir overflow rate, gpd/ft = Flow, gpd / Weir length, ft

Weir overflow rate, gpd/ft = $\frac{2,710,000 \text{ gpd}}{340 \text{ ft}}$ = 7.971 gpd/ft, round to 8,000 gpd/ft

56. **a.** inhibit bacterial growth and replication
Reference: *Water Treatment Operator Training Handbook (AWWA)*

57. **b.** 5,080 gal/ft$^2$
Reference: *Math for Water Treatment Operators (AWWA)*

$$\text{UFRV, gal/ft}^2 = \frac{\text{Total gallons filtered}}{\text{Filter surface area, ft}^2}$$

$$\text{UFRV, gal/ft}^2 = \frac{2,845,200 \text{ gal}}{(28.0 \text{ ft})(20.0 \text{ ft})} = 5,080.7 \text{ gal/ft}^2\text{, round to } 5,080 \text{ gal/ft}^2$$

58. **b.** 65% to 70%
Reference: *Water System Operations (WSO) Series (AWWA)*

59. **c.** adsorption and oxidation
Reference: *Water System Operations (WSO) Series (AWWA)*

60. **b.** Calcium bicarbonate, magnesium carbonate
Reference: *Water System Operations (WSO) Series (AWWA)*

61. **b.** *Cryptosporidium,* does not produce DBPs
Reference: *Water Treatment Operator Training Handbook (AWWA)*

62. **c.** broken diaphragm
Reference: *Water System Operations (WSO) Series (AWWA)*

63. **c.** 10 g of copper sulfate and 10 g of lime
Reference: *Water System Operations (WSO) Series (AWWA)*

64. **b.** Scale deposits or corrosive water
Reference: *Water System Operations (WSO) Series (AWWA)*

65. **c.** 1.1 ft/s

Reference: *Math for Water Treatment Operators (AWWA)*

Equation: $Q\left(\text{Flow, ft}^3/\text{s}\right) = \left(\text{Area, ft}^2\right)(\text{Velocity, ft/s})$

$8.5\ \text{ft}^3/\text{s} = (4.05\ \text{ft})(1.90\ \text{ft})(\text{Velocity, ft/s})$ Now solve for velocity:

$$\text{Velocity, ft/s} = \frac{8.5\ \text{ft}^3/\text{s}}{(4.05\ \text{ft})(1.90\ \text{ft})} = 1.1\ \text{ft/s}$$

66. **d.** chemical and mechanical

Reference: *Water System Operations (WSO) Series (AWWA)*

67. **c.** first customer

Reference: *Water System Operations (WSO) Series (AWWA)*

68. **b.** protect public health

Reference: *Water System Operations (WSO) Series (AWWA)*

69. **c.** calcium carbonate, calcium carbonate

Reference: *Water System Operations (WSO) Series (AWWA)*

70. **a.** Better adsorption of organics

Reference: *Water System Operations (WSO) Series (AWWA)*

71. **d.** All of the above

Reference: *Water System Operations (WSO) Series (AWWA)*

72. **c.** grains/gal

Reference: *Water System Operations (WSO) Series (AWWA)*

73. **a.** 550 $\text{ft}^2$ for each filter

Reference: *Math for Water Treatment Operators (AWWA)*

First, calculate the number of gallons per minute.

$$\left(58\ \text{ft}^3/\text{s}\right)\left(7.48\ \text{gal/ft}^3\right)(60\ \text{s/min}) = 26,030.4\ \text{gpm}$$

$$\text{Filtration rate} = \frac{\text{Flow rate, gpm}}{\text{Filter surface area, ft}^2}$$ Rearranging the equation:

$$\text{Filter surface area, ft}^2 = \frac{\text{Flow rate, gpm}}{\text{Filter surface area, ft}^2}$$

$$\text{Filter surface area, ft}^2 = \frac{26,030.4\ \text{gpm}}{5.87\ \text{gpm/ft}^2} = 434.48\ \text{ft}^2\ \text{(for all 8 filters)}$$

$$\text{Filter area for each filter} = 4,434.48\ \text{ft}^2/8\ \text{filters} = 550\ \text{ft}^2\ \text{for each filter}$$

74. **b.** brown or black

Reference: *Water System Operations (WSO) Series (AWWA)*

75. **b.** lower, resistance to

Reference: *Water System Operations (WSO) Series (AWWA)*

76. **c.** 247 mg/L of $Ca(OH)_2$

Reference: *Math for Water Treatment Operators (AWWA)*

Calculate the hydrated lime required in mg/L. Use an excess lime dosage of 15% (115% or 1.15 in decimal form).

Equation: $\text{Hydrated lime feed, mg/L} = \frac{(A + B + C + D)(1.15 - \text{excess})}{\text{Lime purity}}$

Where A is $CO_2$ in source water: $A = (\text{mg/L as } CO_2)(74.1/44.0)$; where 74.1 = molecular weight (MW) of $Ca(OH)_2$.

Substitute known values and solve:

$A = (9.85\text{ mg/L})(74.1/44.0) = 16.59\text{ mg/L}$

Where B is bicarbonate (total) alkalinity removed in softening:

$B = (\text{mg/L as } CaCO_3 \text{ removed})(74.1/100.1)$ – where 100.1 = MW as $CaCO_3$.

Substitute known values and solve:

$B = (187\text{ mg/L} - 32\text{ mg/L})(74.1/100.1) = 114.74\text{ mg/L}$

Where C is hydroxide alkalinity in softened effluent: $C = (\text{mg/L as } CaCO_3)(74.1/100.1)$.

In this case, there is no hydroxide alkalinity; therefore, C = 0

Where D is magnesium removed in softening: $D = \left(\text{mg/L as } Mg^{2+}\right)(74.1/24.3)$ – where 24.3 = MW of $Mg^{2+}$

$D = (29\text{ mg/L} - 8.6\text{ mg/L})(74.1/24.3) = 62.21\text{ mg/L}$

$$\text{Hydrated lime feed, mg/L} = \frac{(16.59\text{ mg/L} + 114.74\text{ mg/L} + 0 + 62.21\text{ mg/L})(1.15)}{90\%/100\%}$$

Hydrated lime feed, mg/L = 247.30 mg/L, round to 247 mg/L of $Ca(OH)_2$

77. **b.** Monomedium filters

Reference: *Water System Operations (WSO) Series (AWWA)*

78. **a.** Localized corrosion and uniform corrosion

Reference: *Water System Operations (WSO) Series (AWWA)*

79. **c.** alkalinity

Reference: *Water Treatment Operator Training Handbook (AWWA)*

80. **b.** 6%

Reference: *Water System Operations (WSO) Series (AWWA)*

81. **c.** NSF International

Reference: *Water System Operations (WSO) Series (AWWA)*

82. **a.** A cross-connection

Reference: *Water System Operations (WSO) Series (AWWA)*

83. **a.** Residual control

Reference: *Water System Operations (WSO) Series (AWWA)*

84. **c.** inorganic chemicals

Reference: *Water System Operations (WSO) Series (AWWA)*

85. **b.** Before the chlorinator

Reference: *Water System Operations (WSO) Series (AWWA)*

86. **d.** All of the above

Reference: *Water System Operations (WSO) Series (AWWA)*

87. **a.** campgrounds

Reference: *Water System Operations (WSO) Series (AWWA)*

88. **d.** saturators

Reference: *Water System Operations (WSO) Series (AWWA)*

89. **d.** filtration

Reference: *Water System Operations (WSO) Series (AWWA)*

90. **d.** Turn off the last three flocculators

Reference: *Water System Operations (WSO) Series (AWWA)*

91. **c.** 9.4

Reference: *Water System Operations (WSO) Series (AWWA)*

92. **a.** 0.342% of polymer

Reference: *Math for Water Treatment Operators (AWWA)*

Know that there are 16 oz in 1 lb and that 1 gal of water = 8.34 lb

First, convert ounces to pounds.

$$\text{Polymer, lb} = 4.58\,\text{oz} \div 16\,\text{oz/lb} = 0.28625\,\text{lb}$$

Next, calculate the percentage of polymer in the water.

$$\text{Polymer, percentage} = \frac{(0.28625\,\text{lb})\,(100\%)}{0.28625\,\text{lb} + 8.34\,\text{lb/gal}(10.0\,\text{al})} = \frac{(0.28625\,\text{lb})\,(100\%)}{0.28625\,\text{lb} + 83.4\,\text{lb}} =$$

$$\text{Polymer, percentage} = \frac{(0.28625\,\text{lb})\,(100\%)}{83.68625\,\text{lb}} = 0.00342\,(100\%) = 0.342\%\ \text{of polymer}$$

93. **a.** 11,500 lb/month of lime; 7.14 mg/L of lime

Reference: *Water System Operations (WSO) Series (AWWA)*

First, convert grams per minute of lime to pounds per day:

$$\text{Lime, lb/d} = (121\,\text{g/min})\,(1\,\text{lb.}/454\,\text{g})\,(1,440\,\text{min/d}) = 383.789\,\text{lb/d}$$

$$\text{Lime, lb/30} - \text{day month} = (383.789\,\text{lb/d})\,(30\,\text{days/month})$$

$$\text{Lime, lb/30} - \text{day month} = 11,513.67\,\text{lb/30} - \text{day month, round to} 11,500\,\text{lb/month of lime}$$

Next, convert gallons per minute to million gallons per day:

$$(4,475\,\text{gpm})\,(1,440\,\text{min/day})\,(1\text{M}/1,000,000) = 6.444\,\text{mgd}$$

Then find the dosage by using the "pounds" equation and rearranging to solve for dosage.

Equation: Number of lb/d = (mgd) (Dosage, mg/L) (8.34 lb/gal)

$$\text{Dosage, mg/L} = \frac{\text{lb/day of lime}}{(\text{mgd})\,(8.34\,\text{lb/gal})}$$

$$\text{Dosage, mg/L} = \frac{383.789\,\text{lb/d}}{(6.444\,\text{mgd})\,(8.34\,\text{lb/d})} = 7.14\,\text{mg/L of lime}$$

94. **a.** It must be carefully monitored to prevent turbidity breakthrough
Reference: *Water System Operations (WSO) Series (AWWA)*

95. **b.** strong oxidizer
Reference: *Water System Operations (WSO) Series (AWWA)*

96. **d.** Both a and b
Reference: *Water Treatment Operator Training Handbook (AWWA)*

97. **b.** Slow sand filtration
Reference: *Water System Operations (WSO) Series (AWWA)*

98. **a.** inadequate mixing in the rapid-mix or flocculation basins
Reference: *Water System Operations (WSO) Series (AWWA)*

99. **d.** 1,210 gpd/ft$^2$
Reference: *Math for Water Treatment Operators (AWWA)*

First, convert the number of cubic feet per second to gallons per day.

$$\left(17.1\ \text{ft}^3/\text{s}\right)(86,400\ \text{s/d})\left(7.48\ \text{gal/ft}^3\right) = 11,051,251\ \text{gpd}$$

Equation: Surface loading rate = gallons per day (gpd) / number of ft$^2$

$$\text{Surface loading rate} = \frac{11,051,251\ \text{gpd}}{(175\ \text{ft})(52.3\ \text{ft})} = 1,207\ \text{gpd/ft}^2\text{, round to } 1,210\ \text{gpd/ft}^2$$

100. **c.** 18 gpm/ft$^2$
Reference: *Math for Water Treatment Operators (AWWA)*

First, calculate the area of the filter in square feet.

Equation: Number of ft$^2$ = (Length, ft) (Width, ft)

Number of ft$^2$ = (32 ft) (24 ft) = 768 ft$^2$

Now, convert cubic feet per second to gallons per minute.

$$\text{Equation: Number of gpm} = \left(\text{ft}^3/\text{s}\right)(60\ \text{s/min})\left(7.48\ \text{gal/ft}^3\right)$$

$$\text{Number of gpm} = \left(30.5\ \text{ft}^3/\text{s}\right)(60\ \text{s/min})\left(7.48\ \text{gal/ft}^3\right) = 13,688.4\ \text{gpm}$$

Then determine backwash rate.

$$\text{Equation: Backwash rate, gpm/ft}^2 = \frac{\text{Flow, gpm}}{\text{Filter area, ft}^2}$$

$$\text{Backwash rate, gpm/ft}^2 = \frac{13,688.4\ \text{gpm}}{768\ \text{ft}^3}$$

$$\text{Backwash rate, gpm/ft}^2 = 17.82\ \text{gpm/ft}^2\text{, round to } 18\ \text{gpm/ft}^2$$

101. **d.** Inadequate flocculation time
Reference: *Water System Operations (WSO) Series (AWWA)*

102. **c.** Reverse osmosis
Reference: *Water System Operations (WSO) Series (AWWA)*

103. **b.** Leopold filter bottom
Reference: *Water Treatment Operator Training Handbook (AWWA)*

104. **c.** 10.6

Reference: *Water System Operations (WSO) Series (AWWA)*

105. **a.** 29 lb/day of $Na_2SiF_6$

Reference: *Math for Water Treatment Operators (AWWA)*

First, determine how many million gallons per day is being treated.

$$\text{mgd} = (1{,}285\,\text{gpm})\,(1{,}440\,\text{min/d}\,(1\text{M}/1{,}000{,}000) = 1.85\,\text{mgd}$$

Because there is natural fluoride (F) present, subtract the natural from the desired to get the dose required.

$$\text{F Dose required} = 1.20\,\text{mg/LF} - 0.08\,\text{mg/L natural F content} = 1.12\,\text{mg/LF}$$

Write the "pounds" equation with the addition of the percentage purity and fluoride (F) content.

$$\text{Silicofluoride, lb/d} = \frac{(\text{mgd})\,(\text{Dosage, mg/L})\,(8.34\,\text{lb/d})}{(\%\text{purity}/100\%)\,(\%\text{Fcontent}/100\%)}$$

$$\text{Silicofluoride, lb/d} = \frac{(1.85\,\text{mgd})\,(1.12\,\text{mg/L})\,(8.34\,\text{lb/d})}{(98\%\text{purity}/100\%)\,(60.5\%\text{Fcontent}/100\%)} = 29.15\,\text{lb/d, round to 29 lb/d of silicofluoride}$$

106. **c.** With lime, then soda ash

Reference: *Water System Operations (WSO) Series (AWWA)*

107. **b.** nominal ratings, absolute ratings

Reference: *Water System Operations (WSO) Series (AWWA)*

108. **c.** magnesium hydroxide

Reference: *Water System Operations (WSO) Series (AWWA)*

109. **c.** Sodium bicarbonate, increase alkalinity, pH

Reference: *Water System Operations (WSO) Series (AWWA)*

110. **b.** Sand trap

Reference: *Water System Operations (WSO) Series (AWWA)*

111. **a.** sediment and other debris

Reference: *Water System Operations (WSO) Series (AWWA)*

112. **c.** colorless, odorless, heavier than air

Reference: *Water System Operations (WSO) Series (AWWA)*

113. **d.** PAC and chlorine added too close together

Reference: *Water System Operations (WSO) Series (AWWA)*

114. **a.** Lime

Reference: *Water System Operations (WSO) Series (AWWA)*

115. **d.** flocculation, sedimentation, and filtration

Reference: *Water System Operations (WSO) Series (AWWA)*

116. **c.** 4.2 gpm/ft

Reference: *Math for Water Treatment Operators (AWWA)*

Equation: $\text{Calcium hardness as mg/L } CaCO_3 = (2.5)\,(\text{Calcium content in mg/L})$

$$\text{Calcium hardness as mg/L } CaCO_3 = (2.5)\,(22\,\text{mg/L, Calcium})$$

$$\text{Calcium hardness as mg/L} CaCO_3 = 55\,\text{mg/L hardness as } CaCO_3$$

117. **b.** Cover the impoundment

Reference: *Water System Operations (WSO) Series (AWWA)*

118. **a.** Sodium hydroxide, NaOH

Reference: *Water System Operations (WSO) Series (AWWA)*

119. **a.** Lime

Reference: *Water System Operations (WSO) Series (AWWA)*

120. **c.** 61 lb of calcium hypochlorite

Reference: *Water System Operations (WSO) Series (AWWA)*

First, determine the number of MG in the tank with the formula below:

$$\text{MG} = (0.785)\,(\text{Diameter, ft})^2\,(\text{Depth, ft})\,\frac{(7.48\,\text{gal})}{\text{ft}^3}\,\frac{(1\,\text{M})}{1,000,000}$$

$$\text{MG} = (0.785)\,(\text{Diameter, ft})^2\,(\text{Depth, ft})\,\frac{(7.48\,\text{gal})}{\text{ft}^3}\,\frac{(1\,\text{M})}{1,000,000} = 0.191215\,\text{MG}$$

Next, use the "pounds/day" equation, but drop the "day" in this particular problem:

$$\text{Calcium hypochlorite, lb} = (\text{mgd})\,(\text{Dosage, mg/L})\,(8.34\,\text{lb/gal})$$

$$\text{Calcium hypochlorite, lb} = (0.191215\,\text{MG})\,(25\,\text{mg/L})\,(8.34\,\text{lb/gal}) = 39.868\,\text{lb}$$

However, the calcium hypochlorite is only 65.0% pure, so divide by 65.0%/100%

$$\frac{39.868\,\text{lb}}{65.0\%\,\text{available}\,\text{Cl}_2/100\%} = 61.335\,\text{lb, round to 61lb of calcium hypochlorite}$$

121. **b.** Sodium hexametaphosphate

Reference: *Water System Operations (WSO) Series (AWWA)*

122. **b.** Coagulation, flocculation, sedimentation, and filtration

Reference: *Water System Operations (WSO) Series (AWWA)*

123. **b.** Mechanical mixers

Reference: *Water Treatment Operator Training Handbook (AWWA)*

124. **a.** viruses

Reference: *Water System Operations (WSO) Series (AWWA)*

125. **c.** manganese, iron, and hydrogen sulfide

Reference: *Water Treatment Operator Training Handbook (AWWA)*

126. **c.** corrode the stainless-steel mesh

Reference: *Water System Operations (WSO) Series (AWWA)*

127. **a.** $CaSO_4$

Reference: *Water System Operations (WSO) Series (AWWA)*

128. **b.** 4 mg/L

Reference: *Water System Operations (WSO) Series (AWWA)*

129. **a.** Wheeler bottom

Reference: *Water Treatment Operator Training Handbook (AWWA)*

130. **d.** 71 mhp

Reference: *Math for Water Treatment Operators (AWWA)*

First, convert million gallons per day to gallons per minute.

Gpm = (2.42 mgd)(1,000,000/1 M)(1 day/1,440 min) = 1,680.56 gpm

$$\text{Equation: mhp} = \frac{(\text{Flow, gpm})(\text{TH, ft})}{(3,960)(\text{Motor efficiency})(\text{Pump efficiency})}$$

$$\text{mhp} = \frac{(1,680.56\,\text{gpm})(109\,\text{ft})}{(3,960)(88\%/100\%\text{motor efficiency})(74\%/100\%\text{pump efficiency})} = 71\,\text{mhp}$$

131. **d.** 2 days

Reference: *Water System Operations (WSO) Series (AWWA)*

132. **d.** 10–11

Reference: *Water System Operations (WSO) Series (AWWA)*

133. **a.** smaller, head loss, excessive under a heavy load

Reference: *Water System Operations (WSO) Series (AWWA)*

134. **a.** 8 psi

Reference: *Water System Operations (WSO) Series (AWWA)*

135. **c.** False floor system

Reference: *Water System Operations (WSO) Series (AWWA)*

136. **c.** 3.8 ft/s in the 8.0-in. pipe

Reference: *Math for Water Treatment Operators (AWWA)*

Note: Water flowing from a smaller diameter pipe to a larger diameter pipe will slow down when going into the larger pipe. Conversely, water flowing from a larger diameter pipe into a smaller diameter pipe (like the following problem) will speed up. In doing problems such as this, use this as a cross-check to the math).

Flow in the 12.0-in. pipe equals the flow in the 8.0-in. pipe as the flow must remain constant:

$Q_1 = Q_2$

Equation: (Area 1)(Velocity 1) = (Area 2)(Velocity 2)

First, find the diameter for the 8.0-in. and 12.0-in. pipes:

Diameter for 8.0-in. = (8.0 in.)(1 ft/12 in.) = 0.667 ft

Diameter for 12.0-in. = (12.0 in.)(1 ft/12 in.) = 1.0 ft

Then determine the areas of each size pipe: Area = (0.785)(Diameter, ft)$^2$

Area 1 (8.0 in.) = (0.785)(0.667 ft)(0.667 ft) = 0.349 ft$^2$

Area 2 (12.0 in.) = (0.785)(1.0 ft)(1.0 ft) = 0.785 ft$^2$

Last, substitute areas calculated and known velocity in 12-in. pipe.

$$\left(0.349\,\text{ft}^2\right)(\text{x, ft/s}) = \left(0.785\,\text{ft}^2\right)(1.68\,\text{ft/s})$$

Solve for x:

$$\text{x, ft/s} = \frac{(0.785\,\text{ft}^2)(1.68\,\text{ft/s})}{(0.349\,\text{ft}^2)} = 3.8\,\text{ft/s in the 8.0 in. pipe}$$

137. **a.** *Cryptosporidium* oocysts, coagulation and filtration
Reference: *Water System Operations (WSO) Series (AWWA)*

138. **d.** high
Reference: *Water System Operations (WSO) Series (AWWA)*

139. **d.** iron
Reference: *Water System Operations (WSO) Series (AWWA)*

140. **d.** potassium permanganate
Reference: *Water System Operations (WSO) Series (AWWA)*

141. **d.** polyphosphates
Reference: *Water System Operations (WSO) Series (AWWA)*

142. **c.** weak, easily penetrates
Reference: *Basic Microbiology for Drinking Water (AWWA)*

143. **c.** concentration of the chlorine and contact time between chlorine and the water.
Reference: *Water System Operations (WSO) Series (AWWA)*

144. **b.** 0.085 mg/L of polymer aid
Reference: *Math for Water Treatment Operators (AWWA)*

First, determine the number of ml/min for the polymer aid (polymer):

$$\frac{(14\,\text{mL})}{10\,\text{min}} = 1.4\,\text{mL/min}$$

Next, calculate the lb/gal for the polymer:

$$\text{Polymer aid, lb/gal} = (\text{Specific gravity})(8.34\,\text{lb/gal})$$

$$\text{Polymer aid, lb/gal} = (1.25)\,(8.34\,\text{lb/gal}) = 10.425\,\text{lb/gal}$$

Then, find the number of mgd:

$$(5,450\,\text{gpm})\,(1,440\,\text{min/d})\,(1\text{M}/1,000,000) = 7.848\,\text{mgd}$$

Use the dosage equation with conversions added for convenience (dosage/conversion equation):

$$\text{Polymer aid, mg/L} = \frac{(\text{mL/min})\,(1,440\,\text{min/d})\,(\text{lb/gal, polymer})}{(3,785\,\text{mL/gal})\,(\text{mgd})\,(8.34\,\text{lb/gal})}$$

$$\text{Polymer aid, mg/L} = \frac{(1.4\,\text{mL/min})\,(1,440\,\text{min/d})\,(10.425\,\text{lb/gal})}{(3,785\,\text{mL/gal})\,(7.848\,\text{mgd})\,(8.34\,\text{lb/gal})} = 0.085\,\text{mg/L of polymer aid}$$

In water treatment, many polymer aids are usually expressed in parts per billion because they are added and work in such low dosages. Also, higher dosages could "blind" the filters.

Thus, to change milligrams per liter to ppb, multiply by 1,000: $(1,000\text{ppb/mg/L})\,(0.085\text{mg/L}) = 85\text{ppb}$

145. **b.** 1,200 $ft^3$
Reference: *Math for Water Treatment Operators (AWWA)*

First, convert the diameter to feet: $\frac{(16.0\,\text{in.})\,(1\,\text{ft})}{12\,\text{in.}} = 1.333\,\text{ft (Diameter)}$

Then, convert the diameter to the radius: radius $= \text{Diameter}/2 = 1.167\,\text{ft}/2 = 0.6667\,\text{ft (radius)}$

Equation for the volume of a pipe in cubic feet $= \pi r^2\,(\text{Length, ft})$ or $(0.785)\,(\text{D, ft})^2(\text{Length, ft})$ :

Using the first equation, the volume = (3.14) (0.6667 ft) (0.6667 ft) (862 ft)

Volume, $ft^3$ = 1,203 $ft^3$, round to 1,200 $ft^3$

Using the second equation, the volume = (0.785) (1.333 ft) (1.333 ft) (862 ft)

Volume, $ft^3$ = 1,202 $ft^3$, round to 1,200 $ft^3$

146. **a.** $CO_2$

Reference: *Water System Operations (WSO) Series (AWWA)*

147. **d.** Only a and c

Reference: *Water System Operations (WSO) Series (AWWA)*

148. **c.** Less than 1.0 ntu

Reference: *Water System Operations (WSO) Series (AWWA)*

149. **c.** Air binding will not occur

Reference: *Water System Operations (WSO) Series (AWWA)*

150. **c.** Streaming current monitors

Reference: *Water System Operations (WSO) Series (AWWA)*

151. **c.** 6.2–6.8

Reference: *Water Treatment Operator Training Handbook (AWWA)*

152. **c.** a chlorine leak detector, 1 ppm

Reference: *Water System Operations (WSO) Series (AWWA)*

153. **b.** 47 min

Reference: *Math for Water Treatment Operators (AWWA)*

First, determine the number of gallons in 14,100 L.

$$\text{Number of gal} = \frac{14,100\ \text{L}}{3.785\ \text{L/gal}} = 3,725.23\ \text{gal}$$

Then divide the number of gallons by the pumping rate.

Time to pump = 3,725.23 gal/80 gpm = 46.565 min, round to 47 min

Note: (0.565 min)(60 s/min) = 34 s, which is over half a minute, thus we need to round to the next higher minute. The unloading time will be 47 min.

154. **c.** Corrosion of metal pipes

Reference: *Water System Operations (WSO) Series (AWWA)*

155. **c.** weak oxidant, viruses

Reference: *Water Treatment Operator Training Handbook (AWWA)*

156. **c.** 0.2 mg/L

Reference: *Water System Operations (WSO) Series (AWWA)*

157. **a.** fine-bubble diffuser

Reference: *Water System Operations (WSO) Series (AWWA)*

158. **b.** True color, turbidity, apparent color, suspended matter

Reference: *Water System Operations (WSO) Series (AWWA)*

159. **c.** fragile, slowed, doesn't shear
Reference: *Water System Operations (WSO) Series (AWWA)*

160. **b.** deeper, coarser
Reference: *Water System Operations (WSO) Series (AWWA)*

161. **c.** the turbidity
Reference: *Water System Operations (WSO) Series (AWWA)*

162. **a.** chemical oxidants
Reference: *Water System Operations (WSO) Series (AWWA)*

163. **a.** 5 mg/L
Reference: *Water Treatment Operator Training Handbook (AWWA)*

164. **d.** yearly
Reference: *Water Treatment Operator Training Handbook (AWWA)*

165. **b.** Polyphosphates
Reference: *Water System Operations (WSO) Series (AWWA)*

166. **b.** chlorine dioxide, chloramines, ozone
Reference: *Water System Operations (WSO) Series (AWWA)*

167. **a.** polyphosphates and silicates
Reference: *Water System Operations (WSO) Series (AWWA)*

168. **a.** physical, pH
Reference: *Water Treatment Operator Training Handbook (AWWA)*

169. **b.** Sulfuric acid
Reference: *Water System Operations (WSO) Series (AWWA)*

170. **b.** Chloroorganics
Reference: *Water System Operations (WSO) Series (AWWA)*

171. **b.** 70 gal/day of $H_2SiF_6$
Reference: *Math for Water Treatment Operators (AWWA)*

First, determine the required fluoride (F) dose.

F req. = F desired – F in raw water F req. = 1.0 mg/L – 0.12 mg/L = 0.88 mg/L

Next, calculate the number of pounds of fluoride needed using the pounds formula.

F, lb/d = (mg/L) (mgd) (8.34 lb/gal)

F, lb/d = (0.88 mg/L) (15.5 mgd) (8.34 lb/gal) = 113.7576 lb/d

Next, determine the pounds of 79% fluoride ion.

$$\frac{113.7576 \text{ lb/d (F)}}{79\%/100\%} = 143.997 \text{ lb/d of 79\% fluoride ion}$$

$$\frac{(143.997 \text{ lb/d})(100\%)}{21\% \text{ solution}} = 685.7 \text{ lb/d of 21\% fluoride solution}$$

$$H_2SiF_6\text{, gal/d} = \frac{685.7 \text{ lb/d of 21\% solution}}{9.83 \text{ lb/gal}} = 69.756 \text{ gal/d, round to 70 gal/d of } H_2SiF_6$$

172. **c.** Softened water

Reference: *Water System Operations (WSO) Series (AWWA)*

173. **d.** All of the above

Reference: *Water Treatment Operator Training Handbook (AWWA)*

174. **a.** 1.25 lb/gal of NaOCl

Reference: *Math for Water Treatment Operators (AWWA)*

$$\text{NaOCl, lb/gal} = \frac{(\text{Solution percent})(8.34\text{ lb/gal})}{100\%}$$

$$\text{NaOCl, lb/gal} = \frac{(12.5\%)(10.0\text{ lb/gal})}{100\%} = 1.25\text{ lb/gal of NaOCl}$$

175. **d.** Highest hourly flow rate, lowest contactor volume, pH, and temperature

Reference: *Water Treatment Operator Training Handbook (AWWA)*

176. **c.** lamella plates, control the growth of algae

Reference: *Water System Operations (WSO) Series (AWWA)*

177. **a.** Inadequate flocculation time

Reference: *Water System Operations (WSO) Series (AWWA)*

178. **b.** 1.67 ft/s in the 14.0-in. pipe

Reference: *Math for Water Treatment Operators (AWWA)*

Note: Water flowing from a smaller diameter pipe to a larger diameter pipe will slow down when going into the larger pipe. Conversely, water flowing from a larger diameter pipe to a smaller diameter pipe will speed up. In doing problems such as this, use this as a cross-check to the math.

Flow in the 10.0-in. pipe equals the flow in the 14.0-in. pipe as the flow must remain constant:

$Q_1 = Q_2$ Since Q, flow = (Area)(Velocity), it follows that:

$$(\text{Area1})(\text{Velocity1}) = (\text{Area2})(\text{Velocity2})$$

First, find the diameters in feet for the 10.0-in. and 14.0-in. pipes:

$$\text{Diameter for 10.0-in.} = 10.0\text{-in.}(1\text{ft}/12\text{in.}) = 0.833\text{ ft}$$

$$\text{Diameter for 14.0-in.} = 14.0\text{-in.}(1\text{ft}/12\text{in.}) = 1.167\text{ ft}$$

Then determine the areas of each size pipe: $\text{Area} = (0.785)(\text{Diameter, ft})^2$

$$\text{Area 1 (10.0-in.)} = (0.785)(0.833\text{ ft})(0.833\text{ ft}) = 0.545\text{ ft}^2$$

$$\text{Area 2 (14.0-in.)} = (0.785)(1.167\text{ ft})(1.167\text{ ft}) = 1.069\text{ ft}^2$$

Last, substitute areas calculated and known velocity in 10.0-in. pipe.

$$\left(0.545\text{ ft}^2\right)(3.28\text{ ft/s}) = \left(1.069\text{ ft}^2\right)(x\text{, ft/s})$$

Solve for x:

$$x\text{, ft/s} = \frac{(0.545\text{ ft}^2)(3.28\text{ ft/s})}{(1.069\text{ ft}^2)} = 1.672\text{ ft/s, round to 1.67 ft/s in the 14.0-in. pipe}$$

179. **c.** Naturally occurring material that makes up greensand

Reference: *Water Treatment Operator Training Handbook (AWWA)*

180. **c.** 1.13 sp gr

Reference: *Math for Water Treatment Operators (AWWA)*

Know that water has a density of 62.4 lb/ft$^3$. Divide the density of the unknown by the density of water.

Equation: Specific gravity (sp gr) = Density of substance/Density of water

sp gr of unknown substance = $\frac{70.4\ lb/ft^3}{62.4\ lb/ft^3}$ = 1.13 sp gr

181. **d.** 200 mg/L of $Ca(OH)_2$

Reference: *Math for Water Treatment Operators (AWWA)*

Calculate the hydrated lime required in milligrams per liter. Use an excess lime dosage of 15% (115% or 1.15 in decimal form)

Equation: Hydrated lime feed, mg/L = $\frac{(A + B + C + D)(1.15 - excess)}{Lime\ purity}$

Where A is $CO_2$ in source water: A = (mg/L as $CO_2$) (74.1/44.0); where 74.1 = molecular weight (MW) of $Ca(OH)_2$.

Substitute known values and solve:

A = (6.8 mg/L) (74.1/44.0) = 11.45 mg/L

Where B is bicarbonate (total) alkalinity removed in softening:

B = (mg/L as $CaCO_3$ removed) (74.1/100.1) –where 100.1 = MW as $CaCO_3$.

B = (192 mg/L – 38.7 mg/L) (74.1/100.1) = 113.48 mg/L

Where C is hydroxide alkalinity in softened effluent: C = (mg/L as $CaCO_3$) (74.1/100.1).

In this case, there is no hydroxide alkalinity; therefore, C = 0

Where D is magnesium removed in softening: D = (mg/L as $Mg^{2+}$) (74.1/24.3) –where 24.3 = MW of $Mg^{2+}$

D= (19.5 mg/L = 8.8 mg/L)(74.1/24.3) = 32.93 mg/L

Hydrated lime feed, mg/L = $\frac{(11.45\ mg/L + 113.48\ mg/L + 0 + 32.93\ mg/L)(1.15)}{92\%/100\%}$

Hydrated lime feed, mg/L = 197.325 mg/L, round to 200 mg/L of $Ca(OH)_2$

182. **d.** 80 µg/L

Reference: *Water System Operations (WSO) Series (AWWA)*

183. **d.** 20%

Reference: *Water System Operations (WSO) Series (AWWA)*

184. **c.** mechanical straining and adsorption

Reference: *Water System Operations (WSO) Series (AWWA)*

## Laboratory Analysis

1. **b.** Spiked samples to test
   Reference: *Water System Operations (WSO) Series (AWWA)*
2. **a.** mixing chemicals
   Reference: *Water System Operations (WSO) Series (AWWA)*
3. **c.** radium, $3.7 \times 10^{10}$
   Reference: *Water System Operations (WSO) Series (AWWA)*
4. **b.** $OH^{-1}$
   Reference: *Water System Operations (WSO) Series (AWWA)*
5. **c.** lactose or lauryl tryptose, 35.5°C, 24–48 hours
   Reference: *Water Treatment Operator Training Handbook (AWWA)*
6. **b.** number of customers it serves
   Reference: *Water System Operations (WSO) Series (AWWA)*
7. **a.** $CO_2$
   Reference: *Water System Operations (WSO) Series (AWWA)*
8. **b.** 8 hours
   Reference: *Water Treatment Operator Training Handbook (AWWA)*
9. **c.** Porcelain dishes
   Reference: *Water System Operations (WSO) Series (AWWA)*
10. **a.** chromium and cadmium, manganese and iron
   Reference: *Water Treatment Operator Training Handbook (AWWA)*
11. **d.** carbonate and noncarbonate hardness
   Reference: *Water System Operations (WSO) Series (AWWA)*
12. **b.** Coliform bacteria
   Reference: *Water System Operations (WSO) Series (AWWA)*
13. **c.** a measurement of the water's capacity to neutralize an acid
   Reference: *Water System Operations (WSO) Series (AWWA)*
14. **c.** Hold the cap, threads down, while collecting the sample with the other hand
   Reference: *Water System Operations (WSO) Series (AWWA)*
15. **d.** very small colloidal particles
   Reference: *Water System Operations (WSO) Series (AWWA)*
16. **b.** Humic content
   Reference: *Water System Operations (WSO) Series (AWWA)*
17. **b.** polytetrafluoroethylene
   Reference: *Water System Operations (WSO) Series (AWWA)*
18. **c.** from the main stream flow
   Reference: *Water System Operations (WSO) Series (AWWA)*

19. **b.** $Fe(OH)_3$ and $MnO_2$

    Reference: *Water System Operations (WSO) Series (AWWA)*

20. **c.** 127.36 mg/L Ca as $CaCO_3$

    Reference: *Math for Water Treatment Operators (AWWA)*

    Equation: $$\frac{\text{Ca hardness, mg/L as } CaCO_3}{\text{Equivalent weight of } CaCO_3} = \frac{\text{Ca, mg/L}}{\text{Equivalent weight of Ca}}$$

    Substitute known values:

    $$\frac{\text{Ca hardness, mg/L as } CaCO_3}{50.045} = \frac{\text{51 mg/LCa}}{20.04}$$

    Rearrange equation:

    $$\text{Ca hardness, mg/L as } CaCO_3 = \frac{\text{(51 mg/L Ca) (50.045)}}{\text{20.04 Ca}}$$

    Ca hardness, mg/L as $CaCO_3$ = 127.36 mg/L, round to 130 mg/L of Ca hardness as $CaCO_3$

21. **d.** All of the above

    Reference: *Water System Operations (WSO) Series (AWWA)*

22. **d.** All of the above

    Reference: *Water System Operations (WSO) Series (AWWA)*

23. **c.** temperature, colder, higher

    Reference: *Water System Operations (WSO) Series (AWWA)*

24. **a.** configuration of

    Reference: *Water System Operations (WSO) Series (AWWA)*

25. **b.** Coliforms

    Reference: *Water System Operations (WSO) Series (AWWA)*

26. **c.** set of blind standards to

    Reference: *Water System Operations (WSO) Series (AWWA)*

27. **d.** Both b and c

    Reference: *Water System Operations (WSO) Series (AWWA)*

28. **b.** taste, odor, and color; water softener or water heater

    Reference: *Water System Operations (WSO) Series (AWWA)*

29. **c.** Sequestering agents

    Reference: *Water System Operations (WSO) Series (AWWA)*

30. **c.** Lead and Copper Rule

    Reference: *Water System Operations (WSO) Series (AWWA)*

31. **b.** nonphosphate, liquid

    Reference: *Water System Operations (WSO) Series (AWWA)*

32. **d.** Every 24 hours

    Reference: *Water System Operations (WSO) Series (AWWA)*

33. **b.** 20,000 lb/yr of Fe removed

    Reference: *Math for Water Treatment Operators (AWWA)*

First, calculate the iron removal in mg/L.

$(0.46 \text{ mg/L})(82\%/100\%) = (0.46 \text{ mg/L})(0.82) = 0.3772 \text{ mg/L}$

Determine the amount of water in MG produced for the year.

$(17.2 \text{ mgd})(365 \text{ days/year}) = 6,278 \text{ MG/year}$

Next, using the "pounds" equation, solve for the number of pounds/year.

$\text{Fe, lb/year} = (\text{MG/year})(\text{Concentration, mg/L})(8.34 \text{ lb/gal})$

$\text{Fe, lb/year} = (6,278 \text{ MG/year})(0.3772 \text{ mg/L})(8.34 \text{ lb/gal}) =$

$19,750 \text{ lb/year, round to } 20,000 \text{ lb/year of Fe removed}$

34. **c.** Immediately after softening and before the clearwell
    Reference: *Water System Operations (WSO) Series (AWWA)*
35. **d.** Alkalinity
    Reference: *Water System Operations (WSO) Series (AWWA)*
36. **a.** 3 TON
    Reference: *Water System Operations (WSO) Series (AWWA)*
37. **d.** 0.044
    Reference: *Water System Operations (WSO) Series (AWWA)*
38. **c.** sill cocks, faucets with hose threads
    Reference: *Water System Operations (WSO) Series (AWWA)*

## Equipment Operation and Maintenance

1. **b.** torque
   Reference: *Pumps and Pumping (ACR)*
2. **c.** calcium carbonate, HCl
   Reference: *Water System Operations (WSO) Series (AWWA)*
3. **a.** 0.1%
   Reference: *Water System Operations (WSO) Series (AWWA)*
4. **b.** control water leakage along the pump's shaft
   Reference: *Pumps and Pumping (ACR)*
5. **b.** The horsepower equivalent to the watts of electric power supplied to a motor
   Reference: *Water System Operations (WSO) Series (AWWA)*
6. **a.** 400 lb/d
   Reference: *Water System Operations (WSO) Series (AWWA)*
7. **d.** The power supplied to a pump by a motor
   Reference: *Water System Operations (WSO) Series (AWWA)*
8. **b.** sticky floc adsorbs to the media
   Reference: *Water System Operations (WSO) Series (AWWA)*

9. **a.** The portion of power delivered to a pump that is actually used to lift water
   Reference: *Water System Operations (WSO) Series (AWWA)*
10. **b.** 6-in., 10-in.
    Reference: *Water System Operations (WSO) Series (AWWA)*
11. **a.** 2,000 lb
    Reference: *Water System Operations (WSO) Series (AWWA)*
12. **a.** Poor lubrication practices
    Reference: *Pumps and Pumping (ACR)*
13. **c.** 2.0 ft
    Reference: *Water System Operations (WSO) Series (AWWA)*
14. **c.** Mild steel
    Reference: *Water System Operations (WSO) Series (AWWA)*
15. **a.** deionized water
    Reference: *Water System Operations (WSO) Series (AWWA)*
16. **c.** chlorine concentration and contact time
    Reference: *Water System Operations (WSO) Series (AWWA)*
17. **b.** Chlorine Institute
    Reference: *Water System Operations (WSO) Series (AWWA)*
18. **b.** no further tightening can be done on the packing gland
    Reference: *Pumps and Pumping (ACR)*
19. **b.** stable, absence of light, elevated temperatures
    Reference: *Water Treatment Operator Training Handbook (AWWA)*
20. **b.** Monthly
    Reference: *Water System Operations (WSO) Series (AWWA)*
21. **a.** 2 in.
    Reference: *Water System Operations (WSO) Series (AWWA)*
22. **b.** Ultraviolet
    Reference: *Water System Operations (WSO) Series (AWWA)*
23. **d.** increases, decreases
    Reference: *Pumps and Pumping (ACR)*
24. **a.** 1 ppm
    Reference: *Water System Operations (WSO) Series (AWWA)*
25. **c.** 30–60 day supply.
    Reference: *Water System Operations (WSO) Series (AWWA)*
26. **a.** proportional
    Reference: *Water System Operations (WSO) Series (AWWA)*
27. **b.** emergency lights
    Reference: *Water System Operations (WSO) Series (AWWA)*

28. **d.** All of the above
    Reference: *Pumps and Pumping (ACR)*
29. **d.** All of the above
    Reference: *Water System Operations (WSO) Series (AWWA)*
30. **a.** iron
    Reference: *Water System Operations (WSO) Series (AWWA)*
31. **b.** As close to the chemical feeder as possible
    Reference: *Water System Operations (WSO) Series (AWWA)*
32. **b.** Water flow
    Reference: *Water Treatment Operator Training Handbook (AWWA)*
33. **b.** control water leakage
    Reference: *Pumps and Pumping (ACR)*
34. **d.** Both b and c
    Reference: *Pumps and Pumping (ACR)*
35. **d.** calcium and magnesium fluoride
    Reference: *Water System Operations (WSO) Series (AWWA)*
36. **b.** 90°
    Reference: *Pumps and Pumping (ACR)*
37. **d.** All of the above
    Reference: *Water System Operations (WSO) Series (AWWA)*
38. **c.** bridging
    Reference: *Water System Operations (WSO) Series (AWWA)*
39. **c.** chloramines
    Reference: *Water System Operations (WSO) Series (AWWA)*
40. **a.** large solids
    Reference: *Pumps and Pumping (ACR)*
41. **c.** Flow proportional control
    Reference: *Water System Operations (WSO) Series (AWWA)*
42. **d.** Both b and c
    Reference: *Water System Operations (WSO) Series (AWWA)*
43. **c.** greenish-yellow
    Reference: *Water System Operations (WSO) Series (AWWA)*
44. **d.** All of the above
    Reference: *Pumps and Pumping (ACR)*
45. **c.** slinger ring
    Reference: *Pumps and Pumping (ACR)*
46. **b.** PVC
    Reference: *Water System Operations (WSO) Series (AWWA)*

47. **d.** All of the above

Reference: *Pumps and Pumping (ACR)*

48. **c.** with as short a distance between the feeder and the water as possible

Reference: *Water System Operations (WSO) Series (AWWA)*

49. **a.** Chlorine dioxide

Reference: *Water System Operations (WSO) Series (AWWA)*

50. **b.** Meter the chlorine gas safely and accurately

Reference: *Water System Operations (WSO) Series (AWWA)*

51. **a.** asbestos base, graphite

Reference: *Pumps and Pumping (ACR)*

52. **c.** the metal-distortion principle

Reference: *Water Treatment Operator Training Handbook (AWWA)*

53. **a.** Kinetic energy

Reference: *Pumps and Pumping (ACR)*

54. **a.** one to three times per year

Reference: *Water System Operations (WSO) Series (AWWA)*

55. **d.** All of the above

Reference: *Water System Operations (WSO) Series (AWWA)*

56. **b.** 52 bhp

Reference: *Math for Water Treatment Operators (AWWA)*

Equation: Brake hp = (hp)(Motor efficiency)

Brake hp = (60 hp) (86%/100% motor efficiency) = 51.6 bhp, round to 52 bhp

57. **d.** 110 mhp

Reference: *Math for Water Treatment Operators (AWWA)*

$$\text{Equation: Motor hp} = \frac{(\text{whp})}{(\text{Motor efficiency})(\text{Pump efficiency})}$$

$$\text{mhp} = \frac{80\text{ whp}}{(91\%/100\%\text{ motor efficiency})(82\%/100\%\text{ pump efficiency})}$$

$$\text{mhp} = \frac{80\text{ whp}}{(0.91\text{ motor efficiency})(0.82\text{ pump efficiency})}$$

mhp = 107.2 mhp, round to 110 mhp

58. **a.** Closed

Reference: *Pumps and Pumping (ACR)*

59. **d.** Doing all of the above

Reference: *Pumps and Pumping (ACR)*

60. **b.** Centrifugal pump

Reference: *Pumps and Pumping (ACR)*

## Source Water Characteristics

1. **b.** Calcium hypochlorite powder or tablets added as gravel is added
   Reference: *Water System Operations (WSO) Series (AWWA)*
2. **b.** Nitrate, shallow wells, agricultural contamination
   Reference: *Water System Operations (WSO) Series (AWWA)*
3. **d.** All of the above
   Reference: *Water Treatment Operator Training Handbook (AWWA)*
4. **c.** radium
   Reference: *Water System Operations (WSO) Series (AWWA)*
5. **a.** 300 pCi/L
   Reference: *Water System Operations (WSO) Series (AWWA)*
6. **c.** Where soil is sandy
   Reference: *Water System Operations (WSO) Series (AWWA)*
7. **d.** All of the above
   Reference: *Water Treatment Operator Training Handbook (AWWA)*
8. **d.** phosphates
   Reference: *Water System Operations (WSO) Series (AWWA)*
9. **b.** 6.0–8.0
   Reference: *Water System Operations (WSO) Series (AWWA)*
10. **c.** Surface grouting around well
    Reference: *Water System Operations (WSO) Series (AWWA)*
11. **a.** Transpiration
    Reference: *Water System Operations (WSO) Series (AWWA)*
12. **d.** 300 mg/L
    Reference: *Water System Operations (WSO) Series (AWWA)*
13. **b.** They clog intake pipes.
    Reference: *Water System Operations (WSO) Series (AWWA)*
14. **a.** Siltstone
    Reference: *Water System Operations (WSO) Series (AWWA)*
15. **b.** *Acanthamoeba*
    Reference: *Water System Operations (WSO) Series (AWWA)*
16. **b.** Frozen ground
    Reference: *Water System Operations (WSO) Series (AWWA)*
17. **b.** steel tape, tenths and hundredths
    Reference: *Water System Operations (WSO) Series (AWWA)*
18. **d.** Slime-producing bacteria and algae
    Reference: *Water System Operations (WSO) Series (AWWA)*

19. **b.** acidic water devoid of oxygen due to anaerobic activity.
Reference: *Water System Operations (WSO) Series (AWWA)*

20. **d.** All of the above
Reference: *Water System Operations (WSO) Series (AWWA)*

21. **d.** 4.0 logs
Reference: *Water System Operations (WSO) Series (AWWA)*

22. **b.** contaminants
Reference: *Water System Operations (WSO) Series (AWWA)*

23. **c.** Specific-capacity
Reference: *Water System Operations (WSO) Series (AWWA)*

24. **b.** carbon dioxide, carbonic acid, corrosivity
Reference: *Water System Operations (WSO) Series (AWWA)*

25. **c.** Where gypsum is present
Reference: *Water System Operations (WSO) Series (AWWA)*

26. **c.** dibromoacetic acids
Reference: *Water System Operations (WSO) Series (AWWA)*

27. **a.** poorer, confined, do not recharge significantly locally
Reference: *Water System Operations (WSO) Series (AWWA)*

28. **c.** Detailed steps to take if contamination ever occurs
Reference: *Water System Operations (WSO) Series (AWWA)*

29. **b.** humic substances
Reference: *Water System Operations (WSO) Series (AWWA)*

30. **d.** All of the above
Reference: *Water System Operations (WSO) Series (AWWA)*

31. **d.** Hydrogen sulfide gas is produced, and pH decreases.
Reference: *Water System Operations (WSO) Series (AWWA)*

32. **a.** salts
Reference: *Water System Operations (WSO) Series (AWWA)*

33. **b.** nutrients
Reference: *Water System Operations (WSO) Series (AWWA)*

34. **c.** percolation
Reference: *Water System Operations (WSO) Series (AWWA)*

35. **d.** siltation, type of soil, vegetation cover
Reference: *Water System Operations (WSO) Series (AWWA)*

36. **d.** Both b and c
Reference: *Water System Operations (WSO) Series (AWWA)*

37. **d.** properties, distribution; Earth's
Reference: *Water System Operations (WSO) Series (AWWA)*

38. **b.** groundwater springs and seeps to most streams
Reference: *Water System Operations (WSO) Series (AWWA)*

39. **d.** medicinal
Reference: *Water System Operations (WSO) Series (AWWA)*

40. **c.** the water quality is not usually as good as water at intermediate depths
Reference: *Water System Operations (WSO) Series (AWWA)*

41. **d.** All of the above
Reference: *Water System Operations (WSO) Series (AWWA)*

42. **c.** two or more observation wells
Reference: *Water System Operations (WSO) Series (AWWA)*

43. **c.** Some bacteria and protozoa
Reference: *Water System Operations (WSO) Series (AWWA)*

44. **d.** Blue-green algae
Reference: *Water System Operations (WSO) Series (AWWA)*

45. **a.** Chemicals
Reference: *Water System Operations (WSO) Series (AWWA)*

46. **d.** All of the above
Reference: *Water System Operations (WSO) Series (AWWA)*

47. **c.** Determine the length of time of the contamination episode; wait until the spill passes the intake before treating more water
Reference: *Water System Operations (WSO) Series (AWWA)*

48. **a.** 60 mg/L
Reference: *Water System Operations (WSO) Series (AWWA)*

49. **a.** Watershed management
Reference: *Water Treatment Operator Training Handbook (AWWA)*

50. **b.** humic, organics
Reference: *Water System Operations (WSO) Series (AWWA)*

51. **b.** at the base of hills
Reference: *Water System Operations (WSO) Series (AWWA)*

52. **d.** Both b and c
Reference: *Water System Operations (WSO) Series (AWWA)*

53. **b.** a spring
Reference: *Water System Operations (WSO) Series (AWWA)*

54. **b.** 5 ft, usually faster than the cable-tool drilling method
Reference: *Water System Operations (WSO) Series (AWWA)*

55. **c.** drawdown
Reference: *Water System Operations (WSO) Series (AWWA)*

## Security, Safety, Compliance, and Administrative Procedures

1. **a.** float type
   Reference: *Water Treatment Operator Training Handbook (AWWA)*
2. **d.** All of the above
   Reference: *Water Treatment Operator Training Handbook (AWWA)*
3. **b.** NIOSH
   Reference: *Water System Operations (WSO) Series (AWWA)*
4. **b.** Safety goggles
   Reference: *Water System Operations (WSO) Series (AWWA)*
5. **d.** Personal protective equipment
   Reference: *Water Treatment Operator Training Handbook (AWWA)*
6. **a.** average, three, month
   Reference: *Water System Operations (WSO) Series (AWWA)*
7. **c.** granite formations
   Reference: *Water System Operations (WSO) Series (AWWA)*
8. **b.** 3-log removal, *Cryptosporidium*
   Reference: *Water System Operations (WSO) Series (AWWA)*
9. **c.** Hydrogen sulfide
   Reference: *Water System Operations (WSO) Series (AWWA)*
10. **c.** Humic and fulvic acids
    Reference: *Water System Operations (WSO) Series (AWWA)*
11. **c.** in every laboratory
    Reference: *Water System Operations (WSO) Series (AWWA)*
12. **a.** Treatment
    Reference: *Water System Operations (WSO) Series (AWWA)*
13. **a.** in 5-, 10-, and 15-min intervals
    Reference: *Water System Operations (WSO) Series (AWWA)*
14. **d.** Dry: none; liquid: 1–7
    Reference: *Water System Operations (WSO) Series (AWWA)*
15. **a.** chlorite
    Reference: *Water System Operations (WSO) Series (AWWA)*
16. **b.** 5 min
    Reference: *Water System Operations (WSO) Series (AWWA)*
17. **c.** Stage 1 Disinfectants and Disinfection Byproducts Rule
    Reference: *Water System Operations (WSO) Series (AWWA)*
18. **c.** 67%
    Reference: *Water Treatment Operator Training Handbook (AWWA)*

19. **a.** Kit A
Reference: *Water System Operations (WSO) Series (AWWA)*
20. **c.** NSF International
Reference: *Water System Operations (WSO) Series (AWWA)*
21. **b.** Appropriation-permit system
Reference: *Water System Operations (WSO) Series (AWWA)*
22. **a.** zero
Reference: *Water System Operations (WSO) Series (AWWA)*
23. **c.** 10,000
Reference: *Water System Operations (WSO) Series (AWWA)*
24. **c.** Proportional plus reset control
Reference: *Water Treatment Operator Training Handbook (AWWA)*
25. **d.** 4 logs
26. **b.** *Cryptosporidium*
Reference: *Water System Operations (WSO) Series (AWWA)*
27. **a.** 0.8 mg/L
Reference: *Water System Operations (WSO) Series (AWWA)*
28. **d.** 500 mg/L, inorganics, organic
Reference: *Water Treatment Operator Training Handbook (AWWA)*
29. **b.** NSF International
Reference: *Water System Operations (WSO) Series (AWWA)*
30. **d.** Floating proportional control
Reference: *Water Treatment Operator Training Handbook (AWWA)*
31. **c.** ultraviolet light
Reference: *Water System Operations (WSO) Series (AWWA)*
32. **d.** Nitrite
Reference: *Water System Operations (WSO) Series (AWWA)*
33. **d.** dangerous and can even be fatal
Reference: *Water System Operations (WSO) Series (AWWA)*
34. **a.** year, OSHA
Reference: *Water Treatment Operator Training Handbook (AWWA)*
35. **b.** To eliminate injuries
Reference: *Water Treatment Operator Training Handbook (AWWA)*
36. **c.** state regulatory agency
Reference: *Water System Operations (WSO) Series (AWWA)*
37. **a.** liver and kidney damage
Reference: *Water System Operations (WSO) Series (AWWA)*

38. **c.** 0.5, 15-, 10
Reference: *Water System Operations (WSO) Series (AWWA)*

39. **d.** All of the above
Reference: *Water Treatment Operator Training Handbook (AWWA)*

40. **c.** all-purpose fire extinguishers, fire blankets
Reference: *Water System Operations (WSO) Series (AWWA)*

41. **c.** hemolytic anemia
Reference: *Water System Operations (WSO) Series (AWWA)*

42. **d.** Stage 2 Disinfectants and Disinfection Byproducts Rule
Reference: *Water System Operations (WSO) Series (AWWA)*

43. **d.** total coliforms, *E. coli*
Reference: *Basic Microbiology for Drinking Water (AWWA)*

44. **c.** *Cryptosporidium*
Reference: *Water System Operations (WSO) Series (AWWA)*

45. **c.** Radon
Reference: *Water System Operations (WSO) Series (AWWA)*

46. **b.** State and federal regulations
Reference: *Water System Operations (WSO) Series (AWWA)*

47. **a.** Utility manager
Reference: *Water Treatment Operator Training Handbook (AWWA)*

48. **c.** explosive, heavier
Reference: *Water System Operations (WSO) Series (AWWA)*

49. **d.** 10 mg/L, magnesium
Reference: *Water System Operations (WSO) Series (AWWA)*

50. **c.** The operator should have checked the atmosphere of the confined space to declassify it as a non-permit space.
Reference: *Water Treatment Operator Training Handbook (AWWA)*

51. **b.** 4.0 mg/L
Reference: *Water System Operations (WSO) Series (AWWA)*

52. **b.** blue green
Reference: *Water System Operations (WSO) Series (AWWA)*

53. **d.** =0.3, 95, >1
Reference: *Water System Operations (WSO) Series (AWWA)*

54. **d.** five
Reference: *Basic Microbiology for Drinking Water (AWWA)*

55. **b.** 2-log, *Cryptosporidium*
Reference: *Water System Operations (WSO) Series (AWWA)*

56. **d.** Every individual associated with the water utility
Reference: *Water Treatment Operator Training Handbook (AWWA)*

57. **d.** Lead and Copper Rule
Reference: *Water System Operations (WSO) Series (AWWA)*

58. **c.** Hydrogen
Reference: *Water System Operations (WSO) Series (AWWA)*

59. **a.** Causes bone disease
Reference: *Water System Operations (WSO) Series (AWWA)*

60. **b.** 4.0 mg/L total chlorine
Reference: *Water System Operations (WSO) Series (AWWA)*

61. **b.** 83, prohibiting use of lead products to convey drinking water
Reference: *Water System Operations (WSO) Series (AWWA)*

62. **d.** All of the above
Reference: *Water System Operations (WSO) Series (AWWA)*

63. **a.** acute, chronic, boil-water notices
Reference: *Water Treatment Operator Training Handbook (AWWA)*

64. **a.** central nervous system
Reference: *Water System Operations (WSO) Series (AWWA)*

65. **b.** The risk of introducing pathogenic organisms
Reference: *Water System Operations (WSO) Series (AWWA)*

66. **b.** 15 min
Reference: *Water System Operations (WSO) Series (AWWA)*

67. **c.** Wilson's disease
Reference: *Water System Operations (WSO) Series (AWWA)*

68. **b.** sulfur dioxide, photochemical oxidation
Reference: *Water System Operations (WSO) Series (AWWA)*

69. **b.** paralyzes one's sense of smell
Reference: *Water System Operations (WSO) Series (AWWA)*

70. **d.** are easily identified
Reference: *Water System Operations (WSO) Series (AWWA)*

71. **b.** Agricultural
Reference: *Water System Operations (WSO) Series (AWWA)*

72. **b.** disinfect and filter the water
Reference: *Water System Operations (WSO) Series (AWWA)*

73. **c.** Rem
Reference: *Water System Operations (WSO) Series (AWWA)*

74. **c.** 3 years
75. **d.** 2,000 ppm
76. **c.** Interim Enhanced SWTR
    Reference: *Water System Operations (WSO) Series (AWWA)*
77. **d.** Fecal contamination
    Reference: *Water System Operations (WSO) Series (AWWA)*
78. **b.** delivery problems may occur, the chemicals over time lose their effectiveness
    Reference: *Water System Operations (WSO) Series (AWWA)*

# II. Water Treatment, Advanced Answers



## Treatment Process

1. **a.** Sedimentation problems
   Reference: *Water System Operations (WSO) Series (AWWA)*
2. **c.** unscented detergent, odor-free
   Reference: *Water System Operations (WSO) Series (AWWA)*
3. **d.** organophosphates
   Reference: *Water System Operations (WSO) Series (AWWA)*
4. **b.** A flow-restricting orifice plate
   Reference: *Water System Operations (WSO) Series (AWWA)*
5. **d.** distribution and detention time
   Reference: *Water System Operations (WSO) Series (AWWA)*
6. **a.** type of disinfectant used and contact time with that disinfectant
   Reference: *Water System Operations (WSO) Series (AWWA)*
7. **a.** Remove from service and backwash thoroughly
   Reference: *Water Treatment Operator Training Handbook (AWWA)*
8. **a.** 6-in. of anthracite
   Reference: *Water System Operations (WSO) Series (AWWA)*
9. **d.** using parallel plates that direct the water downward then upward at 55°
   Reference: *Water Treatment Operator Training Handbook (AWWA)*
10. **a.** 0.5 ft/min
    Reference: *Water Treatment Operator Training Handbook (AWWA)*
11. **b.** It precipitates iron, which is needed by the algae for chlorophyll production.
    Reference: *Water System Operations (WSO) Series (AWWA)*
12. **a.** Granular activated carbon, rather expensive
    Reference: *Water System Operations (WSO) Series (AWWA)*
13. **a.** Nanofiltration
    Reference: *Water System Operations (WSO) Series (AWWA)*

14. **d.** 18-in. of anthracite
    Reference: *Water Treatment Operator Training Handbook (AWWA)*
15. **b.** It removes $CO_2$, which enhances oxidation because it raises the pH.
    Reference: *Water System Operations (WSO) Series (AWWA)*
16. **b.** phenols
    Reference: *Water System Operations (WSO) Series (AWWA)*
17. **d.** Sulfuric acid
    Reference: *Water System Operations (WSO) Series (AWWA)*
18. **a.** zebra mussels
    Reference: *Water System Operations (WSO) Series (AWWA)*
19. **c.** 2.0-log
    Reference: *Water Treatment Operator Training Handbook (AWWA)*
20. **b.** GAC, anthracite
    Reference: *Water System Operations (WSO) Series (AWWA)*
21. **d.** organic matter, anionic, cation exchange
    Reference: *Water System Operations (WSO) Series (AWWA)*
22. **d.** bacteria
    Reference: *Water System Operations (WSO) Series (AWWA)*
23. **c.** Heat and washing with sodium hydroxide
    Reference: *Water Treatment Operator Training Handbook (AWWA)*
24. **c.** $CaCO_3$
    Reference: *Water System Operations (WSO) Series (AWWA)*
25. **a.** Spiral wound and hollow fiber
    Reference: *Water System Operations (WSO) Series (AWWA)*
26. **a.** flocculation, quickly, previously formed floc particles
    Reference: *Water System Operations (WSO) Series (AWWA)*
27. **b.** helix-type feeder, secondary
    Reference: *Water System Operations (WSO) Series (AWWA)*
28. **b.** 15 min
    Reference: *Water System Operations (WSO) Series (AWWA)*
29. **a.** 1–2
    Reference: *Water System Operations (WSO) Series (AWWA)*
30. **b.** It can be used for both groundwater and surface water.
    Reference: *Water System Operations (WSO) Series (AWWA)*
31. **d.** ozone oxidation
    Reference: *Water System Operations (WSO) Series (AWWA)*

32. **b.** 39.81% Cu

Reference: *Math for Water Treatment Operators (AWWA)*

Equation for calculating the percentage of copper in $CuSO_4$ is $\% Cu = \frac{(\text{Molecular wt of Cu})(100)}{\text{Molecular wt of } CuSO_4}$

First, determine the molecular weight of each of the elements in the compound:

| Element | Number of atoms | | Atomic Wt | | Molecular Wt |
|---|---|---|---|---|---|
| Cu | 1 | x | 63.54 | = | 63.54 |
| S | 1 | x | 32.064 | = | 32.064 |
| O | 4 | x | 15.9994 | = | 63.9976 |
| | | | Molecular weight of $CuSO_4$ | = | 159.6016 |

The molecular weight of Cu in $CuSO_4$ is 63.54. Substituting:

$$\%Cu = \frac{(63.54)(100)}{159.6016} = 39.81\% \text{ Cu}$$

33. **b.** 10 ntu

Reference: *Water Treatment Operator Training Handbook (AWWA)*

34. **d.** Copper sulfate and powdered activated carbon

Reference: *Water System Operations (WSO) Series (AWWA)*

35. **a.** Uneven flow over the weir

Reference: *Water System Operations (WSO) Series (AWWA)*

36. **c.** Ozone

37. **d.** Agglomerated microfloc

Reference: *Water System Operations (WSO) Series (AWWA)*

38. **b.** two, specific gravity

Reference: *Water Treatment Operator Training Handbook (AWWA)*

39. **c.** 5–10 mm

Reference: *Water System Operations (WSO) Series (AWWA)*

40. **b.** Potassium permanganate

Reference: *Water System Operations (WSO) Series (AWWA)*

41. **c.** Makes the floc heavier

Reference: *Water System Operations (WSO) Series (AWWA)*

42. **b.** when the filter media has been drained and the media is still damp

Reference: *Water Treatment Operator Training Handbook (AWWA)*

43. **b.** once per year

Reference: *Water System Operations (WSO) Series (AWWA)*

44. **a.** daily

Reference: *Water System Operations (WSO) Series (AWWA)*

45. **d.** 50%

Reference: *Water Treatment Operator Training Handbook (AWWA)*

46. **d.** Split treatment

Reference: *Water System Operations (WSO) Series (AWWA)*

47. **c.** float, warms, air binding
Reference: *Water System Operations (WSO) Series (AWWA)*

48. **d.** very high heat in the presence of steam, activating
Reference: *Water System Operations (WSO) Series (AWWA)*

49. **c.** chemically bound to the aluminum and ferric hydroxide floc
Reference: *Water System Operations (WSO) Series (AWWA)*

50. **b.** 11, 10.0–10.6
Reference: *Water Treatment Operator Training Handbook (AWWA)*

51. **b.** They work over a wider pH range.
Reference: *Water System Operations (WSO) Series (AWWA)*

52. **a.** adsorption, organic substances
Reference: *Water System Operations (WSO) Series (AWWA)*

53. **c.** lower, acid addition, coagulation
Reference: *Water Treatment Operator Training Handbook (AWWA)*

54. **a.** Lime, alkalinity
Reference: *Water System Operations (WSO) Series (AWWA)*

55. **d.** a manometer
Reference: *Water Treatment Operator Training Handbook (AWWA)*

56. **a.** Sulfuric acid
Reference: *Water System Operations (WSO) Series (AWWA)*

57. **d.** Hydrogen and oxygen
Reference: *Water System Operations (WSO) Series (AWWA)*

58. **c.** 20%
Reference: *Water System Operations (WSO) Series (AWWA)*

59. **b.** Vacuum filters
Reference: *Water System Operations (WSO) Series (AWWA)*

60. **c.** Acids, lower, reverse osmosis membranes
Reference: *Water System Operations (WSO) Series (AWWA)*

61. **d.** 6.0 in.
Reference: *Water System Operations (WSO) Series (AWWA)*

Know : $\text{Flow, ft}^3/\text{s} = \left(\text{Area, ft}^2\right)(\text{Velocity, ft/s})$ ; where the $\text{Area} = (0.785)(\text{Diameter, ft})^2$

$$0.35\ \text{ft}^3/\text{s} = (0.785)(\text{Diameter, ft})^2(1.75\ \text{ft/s})$$

Rearrange and solve for the diameter.

$$D^2, \text{ft} = \frac{0.35\ \text{ft}^3/\text{s}}{(0.785)(1.75\ \text{ft/s})}$$

$$D^2, \text{ft} = 0.255\ \text{ft}$$

Next, take the square root of both sides.

$$\sqrt{D^2}, \text{ft} = \sqrt{0.255}\ \text{ft}$$

$D, ft = 0.505 ft$

Then, convert ft to inches.

$$\text{Diameter, in} = (0.505 \text{ ft})(12 \text{ in/ft}) = 6.06 \text{ in., round to } 6.0 \text{ in.}$$

62. **b.** 1.2 ft/s

Reference: *Math for Water Treatment Operators (AWWA)*

First, convert the number of gallons per minute to cubic feet per second.

$$\text{Number of ft}^3\text{/s} = \frac{182 \text{ gpm}}{(7.48 \text{ gal/ft}^3)(60 \text{ s/min})} = 0.4055 \text{ ft}^3\text{/s}$$

Next, convert the diameter from inches to feet.

$$\text{Pipe diameter, ft} = (8.0 \text{ in.})(1 \text{ ft}/12 \text{ in.}) = 0.667 \text{ ft}$$

Equation: $\text{Flow, ft}^3\text{/s} = (\text{Area, ft}^2)(\text{Velocity, ft/s})$, where the $\text{Area} = (0.785)(\text{Diameter, ft})^2$

$$0.4055 \text{ ft}^3\text{/s} = (0.785)(0.667 \text{ ft})(0.667 \text{ ft})(\text{Flow, ft/s})$$

Rearrange and solve for the flow in fps.

$$\text{Flow, ft/s} = \frac{0.4055 \text{ ft}^3\text{/s}}{(0.785)(0.667 \text{ ft})(0.667 \text{ ft})} = 1.16 \text{ ft/s, round to } 1.2 \text{ ft/s}$$

63. **d.** 2 hydrogen ($H^+$) ions to form hydrogen gas

Reference: *Water System Operations (WSO) Series (AWWA)*

64. **a.** epichlorohydrin

Reference: *Water System Operations (WSO) Series (AWWA)*

65. **d.** Above 10.6

Reference: *Water System Operations (WSO) Series (AWWA)*

66. **b.** daily, monthly

Reference: *Water Treatment Operator Training Handbook (AWWA)*

67. **b.** two, excessive head losses, mudball formation

Reference: *Water Treatment Operator Training Handbook (AWWA)*

68. **c.** Cascade aerators

Reference: *Water System Operations (WSO) Series (AWWA)*

69. **c.** Prevent flow surges

Reference: *Water System Operations (WSO) Series (AWWA)*

70. **d.** divalent metallic cations and minerals dissolved in the water

Reference: *Water System Operations (WSO) Series (AWWA)*

71. **b.** 475 ppb of coagulant aid

Reference: *Water System Operations (WSO) Series (AWWA)*

First, determine the number of mL/min for the coagulant aid:

$$\frac{(112 \text{ mL})}{5 \text{ min}} = 22.4 \text{ mL/min}$$

Use the dosage equation:

$$\text{Coagulant aid, ppb} = \frac{(\text{mL/min})(1,440\ \text{min/day})(\text{lb/gal, Coagulant Aid})(0.50\%)(1,000\ \text{ppb/mg/L})}{(3,785\ \text{mL/gal})(\text{mgd})(8.34\ \text{lb/gal})(100\%)}$$

$$\text{Coagulant aid, mg/L} = \frac{(22.4\ \text{mL/min})(1,440\ \text{min/day})(10.05\ \text{lb/gal})(50\%)(1,000\ \text{ppb/mg/L})}{(3,785\ \text{mL/gal})(10.8\ \text{mgd})(8.34\ \text{lb/gal})(100\%)}$$

$$\text{Coagulant aid, mg/L} = 475\ \text{ppb of coagulant aid}$$

72. **c.** slurry feed system

Reference: *Water System Operations (WSO) Series (AWWA)*

73. **b.** 20 grams of NaOH dissolved in 1 L of deionized water

Reference: *Math for Water Treatment Operators (AWWA)*

$$\text{Normality (N)} = \frac{\text{number of gram} - \text{equivalents of solute}}{\text{number of liters of solution}}$$

$$\text{Substitution: } 0.5\,\text{N} = \frac{\text{number of gram} - \text{equivalents of solute}}{1\text{L of solution}} = 0.5\,\text{gram} - \text{equivalents are required}$$

The formula weight for NaOH = gram equivalent (gr-eq) weight in this case.

NaOH, grams = (Number of gr-eq)(Number g/gr-eq)  Substitute:

(0.5 gr-eq)(40.00 g/gr-eq) = 20 grams of NaOH dissolved in 1 L of deionized water

74. **c.** 90%

Reference: *Water System Operations (WSO) Series (AWWA)*

75. **a.** Nanofiltration and reverse osmosis

Reference: *Water System Operations (WSO) Series (AWWA)*

76. **d.** Sodium hydroxide

Reference: *Water System Operations (WSO) Series (AWWA)*

77. **d.** 35°

Reference: *Water System Operations (WSO) Series (AWWA)*

78. **b.** flux, 99%

Reference: *Water System Operations (WSO) Series (AWWA)*

79. **c.** calcium, such as lime

Reference: *Water Treatment Operator Training Handbook (AWWA)*

80. **a.** Soluble iron

Reference: *Water System Operations (WSO) Series (AWWA)*

81. **b.** conventional filtration treatment, synthetic organic

Reference: *Water System Operations (WSO) Series (AWWA)*

82. **c.** scale, scale

Reference: *Water System Operations (WSO) Series (AWWA)*

83. **b.** Reverse osmosis

Reference: *Water System Operations (WSO) Series (AWWA)*

84. **d.** 589 mL/min of alum

Reference: *Math for Water Treatment Operators (AWWA)*

First, find the pounds per gallon for alum. Alum, lb/gal = (Specific gravity) (8.34 lb/gal)

Alum, lb/gal alum = (1.26) (8.34 lb/gal) = 10.51 lb/gal

Equation for dosage: $\text{Dosage, mg/L} = \dfrac{(\text{mL/min})\,(1{,}440\ \text{min/d})\,(\text{Alum, lb/gal})\,(\%\text{purity})}{(\text{mgd})\,(8.34\ \text{lb/gal})\,(3{,}785\ \text{mL/gal})}$

Rearrange to solve for unknown (milliliters per minute).

$$\text{Alum, mL/min} = \frac{(\text{Dosage, mg/L})\,(\text{mgd})\,(8.34\ \text{lb/gal})\,(3{,}785\ \text{ml/gal})}{(\text{Alum, lb/gal})\,(\%\ \text{purity})\,(1{,}440\ \text{mL/min})}$$

$$\text{Alum, mL/min} = \frac{(8.35\ \text{mg/L})\,(16.4\ \text{mgd})\,(8.34\ \text{lb/gal})\,(3{,}785\ \text{mL/gal})}{(10.51\ \text{lb/gal})\,(48.5\%/100\%\ \text{purity})\,(1{,}440\ \text{min/d})} = 589\text{mL/min of alum}$$

85. **b.** Sodium hexametaphosphate

Reference: *Water System Operations (WSO) Series (AWWA)*

86. **c.** To oxidize iron and manganese

Reference: *Water System Operations (WSO) Series (AWWA)*

87. **a.** sludge thickener, stirring mechanism

Reference: *Water System Operations (WSO) Series (AWWA)*

88. **b.** Turbidity and organics

Reference: *Water System Operations (WSO) Series (AWWA)*

89. **a.** Baffled chambers and static mixers

Reference: *Water System Operations (WSO) Series (AWWA)*

90. **b.** coagulants, fine calcium and magnesium precipitates

Reference: *Water System Operations (WSO) Series (AWWA)*

91. **c.** Chlorine

Reference: *Water System Operations (WSO) Series (AWWA)*

92. **b.** Once per year

Reference: *Water System Operations (WSO) Series (AWWA)*

93. **b.** Magnesium

Reference: *Water Treatment Operator Training Handbook (AWWA)*

94. **d.** pellet reactor, quite rapidly

Reference: *Water System Operations (WSO) Series (AWWA)*

95. **d.** Potassium permanganate

Reference: *Water System Operations (WSO) Series (AWWA)*

96. **d.** 1.67 mL of alum

Reference: *Math for Water Treatment Operators (AWWA)*

First, find the number of pounds per gallon of alum.

Alum, lb/gal = (sp gr)(8.34 lb/gal) = (1.25)(8.34 lb/gal) = 10.425 lb/gal

Next, determine the number of grams per milliliter.

$$\text{Alum, g/mL} = \frac{(10.425\ \text{lb/gal})\,(48.0\%\ \text{purity})\,(454\ \text{grams/lb})}{(3{,}785\ \text{mL/gal})\,(100\%)}$$

Alum, g/mL = 0.60 g/mL

Convert grams per milliliter to milligrams per milliliter.

Alum, mg/mL = (0.60 g/mL)(1,000 mg/g) = 600 mg/mL

Last, determine the number of milliliters required.

Equation: $C_1V_1 = C_2V_2$

(600 mg/mL)(x, mL) = (1 mg/L)(1,000 mL)

x, mL = 1.67 mL of alum

Thus, add 1.67 mL of alum to a 1,000 mL flask and add deionized water to the 1,000 mL mark on the flask.

97. **d.** the floc is composed mostly of magnesium hydroxide particles

Reference: *Water System Operations (WSO) Series (AWWA)*

98. **b.** Nanofiltration and reverse osmosis

Reference: *Water System Operations (WSO) Series (AWWA)*

99. **b.** 1.0 ft/s, 0.9 ft/s to 1.3 ft/s

Reference: *Water Treatment Operator Training Handbook (AWWA)*

100. **d.** spreading to most waters of North America

Reference: *Water System Operations (WSO) Series (AWWA)*

101. **a.** Elevated pH values

Reference: *Water System Operations (WSO) Series (AWWA)*

102. **d.** 1.61 hr

Reference: *Math for Water Treatment Operators (AWWA)*

First, determine the number of gallons in the four flocculation basins and the sedimentation basin.

Equation: Volume, gal = (Length, ft)(Width, ft)(Depth, ft)(7.48 gal/ft$^3$)(4 basins)

$$\text{Floc basins volume gal} = (49.8\,\text{ft})\,(23.9\,\text{ft})\,(10.8\,\text{ft})\left(7.48\,\text{gal/ft}^3\right)(4\text{ basins}) \qquad = 384,603\,\text{gal}$$

$$\text{Volume, gallons in sedimentation basin} = (128\,\text{ft})\,(79.8\,\text{ft})\,(10.1\,\text{ft})\left(7.48\,\text{gal/ft}^3\right) = \frac{771,677\,\text{gal}}{1,156,280\,\text{gal}}$$

Next, convert million gallons per day to gallons per hour.

$$(17.2\ mgd)\,(1\,\text{day}/24\,\text{hours})\,(1,000,000/1\text{M}) = 716,667\,\text{gph}$$

Write the equation with units asked for in the question:

$$\text{Detention time, h} = \frac{\text{Volume, gal}}{\text{Flow rate, gph}}$$

$$\text{Detention time, h} = \frac{1,156,280\,\text{gal}}{716,667\,\text{gph}} = 1.613\,\text{h, round to } 1.61\,\text{h}$$

Note: Detention times are theoretical because basins begin to fill with settled floc and other debris.

103. **a.** through core sampling

Reference: *Water System Operations (WSO) Series (AWWA)*

104. **b.** decrease, decrease

Reference: *Water Treatment Operator Training Handbook (AWWA)*

105. **c.** temperature override device, dangerous temperatures are reached

Reference: *Water System Operations (WSO) Series (AWWA)*

106. **d.** chlorine dioxide

Reference: *Water Treatment Operator Training Handbook (AWWA)*

107. **a.** 9.13 mg/L of alum

Reference: *Math for Water Treatment Operators (AWWA)*

First, determine the number of ml/min for the alum:

$$\frac{985\text{ ml}}{5\text{ min}} = 197\text{ mL/min}$$

Next, find the number of mgd being treated:

$$(3{,}440\text{ gpm})\,(1{,}440\text{ min/day})\,(1\text{M} \div 1{,}000{,}000) = 4.9536\text{ mgd}$$

Use the expanded dosage equation.

$$\text{Equation: Alum, mg/L} = \frac{(\text{mL/min})\,(1{,}440\text{ min/day})\,(\text{Alum, lb/gal})\,(100\%)}{(3{,}785\text{ mL/gal})\,(\text{mgd})\,(8.34\text{ lb/gal})\,(48.5\%\text{Purity})}$$

$$\text{Alum, mg/L} = \frac{(197\text{ mL/min})\,(1{,}440\text{ min/day})\,(10.38\text{ lb/gal})\,(48.5\%)}{(3{,}785\text{ mL/gal})\,(4.9536\text{ mgd})\,(8.34\text{ lb/gal})\,(100\%)} = 9.13\text{ mg/L of alum}$$

108. **d.** >120 mg/L

Reference: *Water Treatment Operator Training Handbook (AWWA)*

109. **d.** All the above

Reference: *Water System Operations (WSO) Series (AWWA)*

110. **a.** Biofouling

Reference: *Water Treatment Operator Training Handbook (AWWA)*

111. **d.** 25

Reference: *Water System Operations (WSO) Series (AWWA)*

112. **a.** 10 NTU

Reference: *Water System Operations (WSO) Series (AWWA)*

113. **c.** 9.6% final solution

Reference: *Math for Water Treatment Operators (AWWA)*

First, find the total volume that would result from mixing these two solutions:

Total volume = 275 gal + 195 gal = 470 gal

Then write the equation: $C_1V_1 + C_2V_2 = C_3V_3$, where $C_1$ and $C_2$ = % concentration of the two solutions before being mixed, $V_1$ and $V_2$ = Volume of the two solutions before being mixed, and $C_3$ and $V_3$ = the resulting % concentration and volume, respectively. Substituting:

$$\frac{(5.8\%)\,(275\text{ gal})}{100\%} + \frac{(15\%)\,(195\text{ gal})}{100\%} = \frac{C_3\,(470\text{ gal})}{100\%}$$

Solving for $C_3$:

$$C_3 = \frac{(15.95\text{ gal} + 29.25\text{ gal})\,(100\%)}{470\text{ gal}}$$

$C_3$= 9.617, round to 9.6% final.

114. **a.** acidified feedwater, unchlorinated product water

Reference: *Water System Operations (WSO) Series (AWWA)*

II. Water Treatment, Advanced Answers

115. **a.** chemical reactions, dissolved, insoluble
Reference: *Water Treatment Operator Training Handbook (AWWA)*

116. **b.** 0.26 mg/L
Reference: *Water System Operations (WSO) Series (AWWA)*

117. **a.** scale formation
Reference: *Water System Operations (WSO) Series (AWWA)*

118. **c.** 15–30 min
Reference: *Water System Operations (WSO) Series (AWWA)*

119. **b.** increases pH and alkalinity
Reference: *Water System Operations (WSO) Series (AWWA)*

120. **d.** All of the above
Reference: *Water System Operations (WSO) Series (AWWA)*

121. **a.** 20% to 25%
Reference: *Water Treatment Operator Training Handbook (AWWA)*

122. **c.** 27,000 lb/yr of Fe removed and 3,600 lb/yr of Mn removed
Reference: *Water System Operations (WSO) Series (AWWA)*

First, calculate the iron removal in ppm.

(0.91 ppm)(89.5%/100%) = (0.91 ppm)(0.895) = 0.8145 ppm or 0.8145 mg/L

Do the same for the manganese.

(0.15 ppm)(71.7%/100%) = (0.15 ppm)(0.717) = 0.1076 ppm or 0.1076 mg/L

Determine the amount of water in MG produced for the year.

(10.88 MGD)(365 days/year) = 3,971.2 MG/year

Next, using the "pounds" equation, solve for the number of pounds/year (yr) for iron and manganese.

$$\text{lb/yr} = (\text{MG/yr})\,(\text{Dosage, mg/L})\,(8.34\ \text{lb/gal})$$

$$\text{Fe, lb/yr} = (3,971.2\ \text{MG/yr})\,(0.8145\ \text{mg/L})\,(8.34\ \text{lb/gal}) = 26,976\ \text{lb/yr, round to } 27,000\ \text{lb/yr Fe removed}$$

$$\text{Mn, lb/yr} = (3,971.2\ \text{MG/yr})\,(0.1076\ \text{mg/L})\,(8.34\ \text{lb/gal}) = 3,564\ \text{lb/yr, round to } 3,600\ \text{lb/yr of Mn removed}$$

123. **b.** High alkalinity
Reference: *Water Treatment Operator Training Handbook (AWWA)*

124. **b.** freeze-thawing beds
Reference: *Water System Operations (WSO) Series (AWWA)*

125. **d.** Remove iron before the exchange resin process
Reference: *Water System Operations (WSO) Series (AWWA)*

126. **b.** Garnet sand, silica sand, and anthracite
Reference: *Water System Operations (WSO) Series (AWWA)*

127. **b.** alkalinity
Reference: *Water Treatment Operator Training Handbook (AWWA)*

128. **a.** deep pitting and tubercles at the anode

Reference: *Water System Operations (WSO) Series (AWWA)*

129. **d.** Polyphosphates

Reference: *Water System Operations (WSO) Series (AWWA)*

130. **b.** Draft-tube aerators

Reference: *Water System Operations (WSO) Series (AWWA)*

131. **a.** Chlorine dioxide

Reference: *Water System Operations (WSO) Series (AWWA)*

132. **d.** Caustic-soda treatment

Reference: *Water System Operations (WSO) Series (AWWA)*

133. **d.** The net ionic charge

Reference: *Water System Operations (WSO) Series (AWWA)*

134. **d.** 10:1 to 20:1.

Reference: *Water System Operations (WSO) Series (AWWA)*

135. **c.** 2–3, 5–7 gpm/ft$^2$

Reference: *Water System Operations (WSO) Series (AWWA)*

136. **d.** 1.87 is the inactivation ratio; yes

Reference: *Math for Water Treatment Operators (AWWA)*

First determine the CT table value.

The CT table value is found by going to the 2.0 log removal table, finding the chart with a pH of 7.4, going down the left column and finding the temperature of 18°C, and then going over to 0.5 mg/L free chlorine residual. Because 0.5 is not shown on the chart, you will need to extrapolate between the numbers that intersect the temperature and the chlorine residuals of 0.4 and 0.6. To extrapolate, subtract the smaller residual from the larger, divide by two, and add this result to the lowest chlorine residual CT table value.

That is: $\text{CT table value} = \frac{35.2 - 34.2}{2} + 34.2 = 34.7 \text{ CT table value}$

Next, calculate the CT.

CT calculated = (Chlorine concentration, mg/L)(Time, min)

CT calculated = (0.5, mg/L)(130 min) = 65 is CT calculated

Next, calculate the inactivation ratio.

$$\text{Inactivation ratio} = \frac{\text{Calculated CT value}}{\text{CT table value}}$$

$$\text{Inactivation ratio} = \frac{65}{34.7} = 1.87 \text{ is the inactivation ratio}$$

Because the inactivation ratio value is greater than 1.0, this system meets the CT criteria and is in compliance.

II. Water Treatment, Advanced Answers

137. **b.** 1.36 mg/L of polymer

Reference: *Math for Water Treatment Operators (AWWA)*

First, determine the number of milliliters per minute for the polymer:

$$\frac{(343\text{ mL})}{10\text{ min}} = 34.3\text{ mL/min}$$

Next, calculate the pounds per gallon for the polymer:

Polymer, lb/gal = (Specific Gravity) (8.34 lb/gal)

Polymer, lb/gal = (1.27) (8.34 lb/gal) = 10.592 lb/gal

Then, find the number of mgd:

(8,470 gpm) (1,440 min/day) (1M/1,000,000) = 12.197 mgd

Use the dosage equation with conversions added for convenience (dosage/conversion equation):

$$\text{Polymer, mg/L} = \frac{(34.3\text{ mL/min})(1,440\text{ min/d})(10.592\text{ lb/gal})}{(3,785\text{ mL/gal})(12.197\text{ mgd})(8.34\text{ lb/gal})} = 1.36\text{ mg/L of polymer}$$

138. **c.** Head loss

Reference: *Water System Operations (WSO) Series (AWWA)*

139. **c.** 0 bicarbonate alkalinity 52 mg/L as carbonate alkalinity; 6 mg/L as hydroxide alkalinity

Reference: *Math for Water Treatment Operators (AWWA)*

First, find the relationship between P alkalinity and T alkalinity by dividing the total alkalinity by 2.

$$\frac{58\text{ mg/L, T alkalinity}}{2} = 29$$

Because the P alkalinity is greater than ½ the T alkalinity, the fourth row in the table above will be used to find the bicarbonate, carbonate, and hydroxide alkalinities.

Bicarbonate alkalinity = 0 from table above

Carbonate alkalinity = 2T – 2P Substitute titration results:

Carbonate alkalinity = 2 (58 mg/L) –2 (32 mg/L) = 116 mg/L – 64 mg/L = 52 mg/L as carbonate alkalinity

Hydroxide alkalinity = 2P – T Again, substitute titration results:

Hydroxide alkalinity = 2(32 mg/L) – 58 mg/L = 64 mg/L – 58 mg/L = 6 mg/L as hydroxide alkalinity

140. **b.** 10%

Reference: *Water System Operations (WSO) Series (AWWA)*

141. **b.** 200 mL

Reference: *Water System Operations (WSO) Series (AWWA)*

142. **a.** installing a GAC contactor

Reference: *Water System Operations (WSO) Series (AWWA)*

143. **b.** Weir overflow rate

Reference: *Water System Operations (WSO) Series (AWWA)*

144. **c.** When there is a noticeable pressure drop between inlet and outlet

Reference: *Water System Operations (WSO) Series (AWWA)*

145. **c.** corrosion- and erosion-resistant, fiberglass

Reference: *Water System Operations (WSO) Series (AWWA)*

146. **c.** filtration, *Giardia, Cryptosporidium*

Reference: *Water Treatment Operator Training Handbook (AWWA)*

147. **d.** Deposits will build up and block the pipes to the sewage plant.

Reference: *Water System Operations (WSO) Series (AWWA)*

148. **d.** synthetic organic chemicals

Reference: *Water System Operations (WSO) Series (AWWA)*

149. **a.** Granular activated carbon

Reference: *Water System Operations (WSO) Series (AWWA)*

150. **a.** filter aids

Reference: *Water System Operations (WSO) Series (AWWA)*

151. **a.** 5.18 psi

Reference: *Math for Water Treatment Operators (AWWA)*

Equation: $\text{psi} = \frac{\text{Depth, ft}}{2.31\text{ ft/psi}}$

$\text{psi} = \frac{9.66\text{ ft}}{2.31\text{ ft/psi}} = 4.182\text{ psi}$ Again, this would be the psi if it were water.

Next, find the specific gravity (sp gr) of the polymer.

$\text{sp gr} = \frac{10.34\text{ lb/gal for alum}}{8.34\text{ lb/gal for water}} = 1.2398\text{ sp gr}$

Then, multiply the psi by the sp gr to determine the psi.

$\text{psi} = (4.182\text{ psi})(1.2398\text{ sp gr}) = 5.18\text{ psi}$

152. **c.** 10.4% final solution

Reference: *Math for Water Treatment Operators (AWWA)*

First, find the total volume that would result from mixing these two solutions:

Total volume = 275 gal + 110 gal = 385 gal

Another way to solve mixture problems is with the following equation:

(Concentration$_1$)(Volume$_1$) + (Concentration$_2$)(Volume$_2$) = (Concentration$_3$)(Volume$_3$), condensed as $C_1V_1 + C_2V_2 = C_3V_3$, where $C_1$ and $C_2$ = % concentration of the two solutions before being mixed, $V_1$ and $V_2$ = Volume of the two solutions before being mixed, and $C_3$ and $V_3$ = the resulting % concentration and volume, respectively. Substituting:

$$\frac{(12.5\%)(275\text{ gal})}{100\%} + \frac{(5.2\%)(110\text{ gal})}{100\%} = \frac{C_3(385\text{ gal})}{100\%}$$

$$34.375\text{ gal} + 5.72\text{ gal} = \frac{C_3(385\text{ gal})}{100\%}$$

Solving for $C_3$:

$$C_3 = \frac{(34.375\text{ gal} + 5.72\text{ gal})(100\%)}{385\text{ gal}} = \frac{(40.095\text{ gal})(100\%)}{385\text{ gal}}$$

$C_3$ = 10.414, round to 10.4% final solution

153. **b.** 1.0, 2.0

Reference: *Water Treatment Operator Training Handbook (AWWA)*

II. Water Treatment, Advanced Answers

154. **a.** 5 mg/L
Reference: *Water System Operations (WSO) Series (AWWA)*

155. **d.** high, two-stage
Reference: *Water Treatment Operator Training Handbook (AWWA)*

156. **b.** $H_2S$
Reference: *Water System Operations (WSO) Series (AWWA)*

157. **c.** Limestone contactors
Reference: *Water System Operations (WSO) Series (AWWA)*

158. **c.** An insoluble silica precipitate often forms.
Reference: *Water System Operations (WSO) Series (AWWA)*

159. **c.** Mixing speed and sludge blanket
Reference: *Water System Operations (WSO) Series (AWWA)*

160. **a.** Proportional-level equal rate control
Reference: *Water Treatment Operator Training Handbook (AWWA)*

161. **d.** pH, temperature, and dissolved oxygen
Reference: *Water System Operations (WSO) Series (AWWA)*

162. **d.** All of the above
Reference: *Water System Operations (WSO) Series (AWWA)*

163. **b.** to keep too many units in service
Reference: *Water Treatment Operator Training Handbook (AWWA)*

164. **d.** downward, at a higher flow rate
Reference: *Water System Operations (WSO) Series (AWWA)*

165. **c.** Ozone
Reference: *Water System Operations (WSO) Series (AWWA)*

166. **c.** color
Reference: *Water Treatment Operator Training Handbook (AWWA)*

167. **c.** fouling
Reference: *Water System Operations (WSO) Series (AWWA)*

168. **b.** 75 mg/L
Reference: *Water System Operations (WSO) Series (AWWA)*

169. **b.** 3–5 ft
Reference: *Water System Operations (WSO) Series (AWWA)*

170. **b.** top of the gravel bed, $^3/_8$-in.
Reference: *Water Treatment Operator Training Handbook (AWWA)*

171. **c.** noncarbonate, recarbonation, lower the pH
Reference: *Water System Operations (WSO) Series (AWWA)*

172. **d.** 0.76 mg/L of caustic

Reference: *Water System Operations (WSO) Series (AWWA)*

First, find number of lb/gal for the caustic.

$(1.278)\,(8.34\text{ lb/gal}) = 10.66\text{ lb/gal}$

Next, convert $ft^3/s$ to mgd:

$\text{Number of mgd} = \left(12.8\text{ ft}^3/\text{s}\right)(86,400\text{ sec/day})\left(7.48\text{ gal/ft}^3\right)(1\text{ M}/1,000,000) = 8.27\text{ mgd}$

Now, find the number of lb/day of caustic usage.

$(52\text{ mL/min})(1,440\text{ min/day})(1\text{ gal}/3,785\text{ mL})(10.66\text{ lb/gal}) = 210.89\text{ lb/day}$

Now, determine the dosage.

$$\text{Dosage, mg/L} = \frac{\text{lb/day}}{(\text{mgd})\,(8.34\text{ lb/gal})}$$

$$\text{Dosage, mg/L} = \frac{(210.89\text{ lb/day})\,(25\%)}{(8.27\text{ mgd})\,(8.34\text{ lb/gal})\,(100\%)} = 0.76\text{ mg/L of caustic}$$

173. **c.** Ozone

Reference: *Water System Operations (WSO) Series (AWWA)*

174. **d.** *Cryptosporidium*

Reference: *Water System Operations (WSO) Series (AWWA)*

175. **d.** Coagulants and chlorine

Reference: *Water System Operations (WSO) Series (AWWA)*

176. **d.** 5.0 ft

Reference: *Water System Operations (WSO) Series (AWWA)*

177. **c.** increases, increases

Reference: *Water System Operations (WSO) Series (AWWA)*

178. **d.** ineffective, thousands of times higher

Reference: *Water System Operations (WSO) Series (AWWA)*

179. **b.** humic acid

Reference: *Water System Operations (WSO) Series (AWWA)*

180. **a.** Use a two-tank setup to eliminate the lime sludge being sucked in from the bottom of the first tank

Reference: *Water System Operations (WSO) Series (AWWA)*

181. **a.** bacteria

Reference: *Water System Operations (WSO) Series (AWWA)*

182. **a.** 18 hr and 34 min

Reference: *Math for Water Treatment Operators (AWWA)*

First, determine the pumping rate in gallons per minute.

$$\text{Pumping rate, gpm} = \frac{(60.5\text{ hp})\,(3,960)\,(81.3\%)}{(385\text{ ft})\,(100\%)} = 505.9\text{ gpm}$$

Next, determine the water level in feet that is in the storage tank.

Equation:

$$\text{Current process reading} = \frac{(\text{Live signal, mA} - \text{4mA offset})(\text{Maximum capacity})}{\text{16 mA span}}$$

Substitute known values and solve for current process reading—in this case, the storage tank's level:

$$\text{Storage tank level, ft} = \frac{(\text{16.07 mA} - \text{4mA offset})(\text{38.5 ft maximum level})}{\text{16 mA span}}$$

Storage tank level = 29.04 ft

Because we need to remove all but 10 ft, it follows that:

Number of feet to remove = 29.04 − 10.0 ft = 19.04 ft

Next, calculate the number of gallons in 19.04 ft in this tank.

Number of gal = $\pi$(Radius, ft)$^2$(Footage to be removed, ft)(7.48 gal/ft$^3$)

Storage tank, gal = $\pi$(35.5 ft)$^2$(19.04 ft)(7.48 gal/ft$^3$)

Storage tank, gal = 3.14(35.5 ft)(35.5 ft)(19.04 ft)(7.48 gal/ft$^3$) = 563,579

Finally, determine how long this will take in hours and minutes.

$$\text{Pumping time, minutes} = \frac{563,579\text{ gal}}{505.9\text{ gpm}} = 1,114\text{ min}$$

$$\frac{1,114\text{ min}}{60\text{ min/h}} = 18.567\text{ hours}$$

Number of minutes in 0.567 hours:

Number of minutes = 0.567(60 h/min) = 34 min

Thus, it will take 18 hr and 34 min

183. **c.** 9.5

Reference: *Water System Operations (WSO) Series (AWWA)*

184. **b.** van der Waals forces

Reference: *Water System Operations (WSO) Series (AWWA)*

185. **d.** Every 5 years

Reference: *Water System Operations (WSO) Series (AWWA)*

186. **d.** dissolving, decrease

Reference: *Water Treatment Operator Training Handbook (AWWA)*

187. **a.** HOCl, chlorine residuals

Reference: *Water System Operations (WSO) Series (AWWA)*

188. **b.** Overcome temperature increases and drops

Reference: *Water System Operations (WSO) Series (AWWA)*

189. **b.** increases as alkalinity decreases

Reference: *Water System Operations (WSO) Series (AWWA)*

190. **b.** Reverse osmosis

Reference: *Water Treatment Operator Training Handbook (AWWA)*

191. **c.** vitrified clay blocks

Reference: *Water System Operations (WSO) Series (AWWA)*

192. **c.** flow fluctuations, channels

Reference: *Water System Operations (WSO) Series (AWWA)*

193. **a.** Lamella plates

Reference: *Water System Operations (WSO) Series (AWWA)*

194. **a.** iron, manganese, IR

Reference: *Water Treatment Operator Training Handbook (AWWA)*

195. **c.** dry ice or liquid $CO_2$, smaller

Reference: *Water System Operations (WSO) Series (AWWA)*

196. **b.** decrease, increases, decreases

Reference: *Water System Operations (WSO) Series (AWWA)*

197. **c.** $CO_2$

Reference: *Water System Operations (WSO) Series (AWWA)*

198. **c.** desorption

Reference: *Water System Operations (WSO) Series (AWWA)*

199. **d.** Chlorine concentration multiplied by the time of contact

Reference: *Water System Operations (WSO) Series (AWWA)*

200. **a.** 420 gal of the 12.5% NaOCl solution are needed

Reference: *Water System Operations (WSO) Series (AWWA)*

Equation: $C_1V_1 + C_2V_2 = C_3V_3$

Where: $C_1$ is the concentration and $V_1$ is the volume of the 12.5% NaOCl solution, $C_2$ is the concentration and $V_2$ is the volume of the water used to dilute the NaOCl solution, and $C_3$ is the final concentration of the mixture and $V_3$ is the final volume of the mixture. Note: Since the NaOCl is being diluted with water, the $C_2$ will be zero since it does not contain any NaOCl.

Since $V_1$ and $V_2$ are unknown, set $V_1 = V_3 - V_2$, since Volume 1 + Volume 2 has to equal Volume 3; that is, $V_1 + V_2 = V_3$, subtracting $V_2$ from both sides of the equation gives $V_1 = V_3 - V_2$

Substituting:

$(12.5\%/100\%)(1{,}000 \text{ gal} - V_2) + (0\%)(V_2) = (5.25\%/100\%)(1{,}000 \text{ gal})$

$125 \text{ gal} - 0.125\, V_2 + 0 = 52.5 \text{ gal}$

Subtract 52.5 gal from both sides of the equation:

$72.5 \text{ gal} - 0.125\, V_2 = 0$

Subtract 72.5 gal from both sides of the equation:

$-0.125\, V_2 = -72.5 \text{ gal}$

Multiply both sides of the equation by (−1):

$0.125\ V_2 = 72.5$ gal

Divide both sides of the equation by 0.125:

$V_2$ = 580 gal of water will be needed to dilute the NaOCl solution of 12.5% to 5.25%

$V_1 = V_3 - V_2$, V1 = 1,000 gal – 580 gal = 420 gal of the 12.5% NaOCl solution are needed.

201. **b.** 2.58 mg/L of polymer

Reference: *Math for Water Treatment Operators (AWWA)*

First, determine the number of milliliters per minute for the polymer: $\frac{(232\text{ mL})}{5\text{ min}} = 46.4\text{ mL/min}$

Next, calculate the pounds per gallon for the polymer:

$$\text{Polymer, lb/gal} = (\text{Specific gravity})(8.34\text{ lb/gal})$$

$$\text{Polymer, lb/gal} = (1.28\text{ sp gr})(8.34\text{ lb/gal}) = 10.675\text{ lb/gal}$$

Then, find the number of million gallons per day:

$$(6,080\text{ gpm})(1,440\text{ min/day})(1\text{M}/1,000,000) = 8.755\text{ mgd}$$

$$\text{Dosage/conversion equation : Polymer, mg/L} = \frac{(\text{mL/min})(1,440\text{ min/d})(\text{Polymer, lb/gal})}{(3,785\text{ mL/gal})(\text{mgd})(8.34\text{ lb/gal})}$$

$$\text{Polymer, mg/L} = \frac{(46.4\text{ mL/min})(1,440\text{ min/d})(10.675\text{ lb/gal})}{(3,785\text{ mL/gal})(8.755\text{ mgd})(8.34\text{ lb/gal})} = 2.58\text{ mg/L of polymer}$$

202. **a.** the attraction of sticky particles to the anthracite or sand and then those attached particles attracting even more sticky particles until the filter is ripened

203. **d.** All of the above

Reference: *Water System Operations (WSO) Series (AWWA)*

204. **a.** 3,000 lb/d of sludge

Reference: *Math for Water Treatment Operators (AWWA)*

Equation: $\text{Sludge} = (8.34\text{ lb/gal})(\text{mgd})\{(2.9\text{ Fe}) + \text{SS} + \text{A}\}$

Where SS = source water suspended solids in mg/L, and A = is other solid treatment chemicals.

First, convert turbidity in ntu to mg/L.

Equation: SS = (B value)(Turbidity, mg/L)

SS = (0.9)(11 ntu) = 9.9

Next, solve for predicted amount of daily sludge produced by substituting known values in the equation above.

$$\text{Sludge} = (8.34\text{ lb/gal})(6.7\text{ mgd})\{(2.9)(14\text{ mg/L}) + 20.4\text{ mg/L} + 3\text{ mg/L})\}$$

$$\text{Sludge} = (8.34\text{ lb/gal})(8.3\text{ mgd})(40.6\text{ mg/L} + 9.9\text{ mg/L} + 5\text{ mg/L})$$

$$\text{Sludge} = (8.34\text{ lb/gal})(6.7\text{ mgd})(55.5\text{ mg/L})$$

$$\text{Sludge} = 3,101\text{ lb/d of sludge}$$

Round to 3,000 lb/d of sludge (because question has one variable with only one significant figure)

205. **a.** pH 4.5

Reference: *Water System Operations (WSO) Series (AWWA)*

206. **a.** 5.5

Reference: *Water Treatment Operator Training Handbook (AWWA)*

207. **d.** exchange capacity, porosity

Reference: *Water System Operations (WSO) Series (AWWA)*

208. **b.** 2.20 mg/L of polymer

Reference: *Water System Operations (WSO) Series (AWWA)*

First, find number of lb/gal for the polymer.

$(1.26\ g/mL)\ (3,785\ mL/gal)\ (1\ lb/454\ g) = 10.50\ lb/gal$

Next, convert $ft^3/s$ to mgd:

$\text{Number of mgd} = \left(30.1 ft^3/s\right)\ (86,400\ sec/day)\ \left(7.48\ gal/ft^3\right)\ (1\ M/1,000,000) = 19.45\ mgd$

Now, find the number of lb/d of polymer usage.

$(89.3\ mL/min)\ (1,440\ min/day)\ (1\ gal/3,785\ mL)\ (10.50\ lb/gal) = 356.73\ lb/d$

Now, determine the dosage.

$$\text{Dosage, mg/L} = \frac{lb/d}{(mgd)\ (8.34\ lb/gal)}$$

$$\text{Dosage, mg/L} = \frac{356.73\ lb/d}{(19.45\ mgd)\ (8.34\ lb/gal)} = 2.199\ mg/L\text{, round to 2.20 mg/L of polymer}$$

209. **c.** 0.5 mg/L

Reference: *Water Treatment Operator Training Handbook (AWWA)*

210. **a.** coagulants, upstream

Reference: *Water System Operations (WSO) Series (AWWA)*

211. **b.** Bromate and formaldehyde production

Reference: *Water Treatment Operator Training Handbook (AWWA)*

212. **c.** polynuclear aromatic hydrocarbons

Reference: *Water System Operations (WSO) Series (AWWA)*

213. **d.** 414 mg/L of $Ca(OH)_2$; 48 mg/L of soda ash

Reference: *Math for Water Treatment Operators (AWWA)*

Calculate the hydrated lime required in milligrams per liter. Use an excess lime dosage of 15% (115% or 1.15 in decimal form).

$$\text{Equation: Hydrated lime feed, mg/L} = \frac{(A + B + C + D)\ (1.15 - \text{excess})}{\text{Lime purity}}$$

Where A is $CO_2$ in source water: A = (mg/L as $CO_2$)(74.1/44), where 74.1 = molecular weight (MW) of $Ca(OH)_2$. Substitute known values and solve:

$A = (25\ mg/L)\ (74.1/44) = 42.10\ mg/L$

Where B is bicarbonate (total) alkalinity removed in softening:

B = (mg/L as $CaCO_3$ removed)(74.1/100.1), where 100.1 = MW as $CaCO_3$.

II. Water Treatment, Advanced Answers

Substitute known values and solve:

$B = (286\ mg/L - 76\ mg/L)(74.1/100.1) = 210\ mg/L$

Where C is hydroxide alkalinity in softened effluent: $C = (mg/L\ as\ CaCO_3)(74.1/100.1)$.

In this case, there is no hydroxide alkalinity; therefore, C = 0

Where D is magnesium (Mg) removed in softening: D = (mg/L as $Mg^{2+}$)(74.1/24.3), where 24.3 = MW of $Mg^{2+}$.

$D = (33\ mg/L - 6.5\ mg/L)(74.1/24.3) = 80.81\ mg/L$

$$\text{Hydrated lime feed, mg/L} = \frac{(42.10\ mg/L + 210\ mg/L + 0 + 80.81\ mg/L)(1.15)}{92.5\%/100\%} = 414 mg/L, Ca(OH)_2$$

Calculate the soda ash required in milligrams per liter. First, find the total hardness removed.

Total hardness removed, mg/L as $CaCO_3$ = Total hardness, mg/L as $CaCO_3$ – Total hardness remaining, mg/L as $CaCO_3$.

Total hardness removed, mg/L as $CaCO_3$ = 412 mg/L – 81 mg/L = 331 mg/L

Noncarbonate hardness, mg/L as $CaCO_3$ = Total hardness removed, mg/L as $CaCO_3$ – Carbonate hardness, m/L as $CaCO_3$.

Noncarbonate hardness, mg/L as $CaCO_3$ = 331 mg/L – 286 mg/L = 45 mg/L as $CaCO_3$.

Soda ash feed, mg/L = (Noncarbonate hardness, mg/L as $CaCO_3$)(106/100.1), where 106 = MW soda ash.

Soda ash feed, mg/L = (45 mg/L)(106/100.1) = 48 mg/L, soda ash

214. **b.** 5–8 gpm/$ft^2$

Reference: *Water System Operations (WSO) Series (AWWA)*

215. **d.** 5.46 mg/L of Lime

Reference: *Math for Water Treatment Operators (AWWA)*

First, find the pounds per day of lime usage. Write the equation.

Lime, lb/d = (g/min)(1,440 min/d)(1 lb/454 g)

Lime, lb/d = (100.2 g/min)(1,440 min/d)(1lb/454g) = 317.8lb/d of lime

Next, convert 10.8 $ft^3$/s to the number of mgd.

$\text{Number of mgd} = (10.8\ ft^3/s)(86,400\ s/d)(7.48\ gal/ft^3)(1M/1,000,000) = 6.98\ mgd$

As in the above problem, rearrange and solve for dosage in milligrams per liter.

$$\text{Dosage, mg/L} = \frac{317.8\ lb/d}{(6.98\ mgd)(8.34\ lb/gal)} = 5.46\ mg/L \text{ of lime}$$

216. **d.** Dichloroamine

Reference: *Water System Operations (WSO) Series (AWWA)*

217. **b.** 8.32 mg/L of soda ash

Reference: *Math for Water Treatment Operators (AWWA)*

First, find the lb/d of soda ash usage.

Soda ash, lb/d = (g/min)(1,440 min/d)(1lb/454 g) = lb/d

Soda ash, lb/d = (162.8 g/min)(1,440 min/d)(1 lb/454 g) = 516 lb/d of soda ash

Then, using the "pounds" equation, calculate the dosage in milligrams per liter.

$$\text{Dosage, mg/L} = \frac{516 \text{ lb/d}}{(7.44 \text{ mgd})(8.34 \text{ lb/gal})} = 8.32 \text{ mg/L of soda ash}$$

## Laboratory Analysis

1. **b.** above pH 4.3, lower pH ranges
   Reference: *Water System Operations (WSO) Series (AWWA)*

2. **a.** Alkalinity caused by $CO_3^{-2}$
   Reference: *Water System Operations (WSO) Series (AWWA)*

3. **c.** color
   Reference: *Water System Operations (WSO) Series (AWWA)*

4. **b.** 2.0–2.5
   Reference: *Water System Operations (WSO) Series (AWWA)*

5. **b.** Sodium hypochlorite
   Reference: *Water System Operations (WSO) Series (AWWA)*

6. **c.** dissolved solid
   Reference: *Water System Operations (WSO) Series (AWWA)*

7. **a.** 3 years
   Reference: *Water System Operations (WSO) Series (AWWA)*

8. **b.** 4.3, methyl orange
   Reference: *Water System Operations (WSO) Series (AWWA)*

9. **d.** viruses, adequate $C \times T$
   Reference: *Water System Operations (WSO) Series (AWWA)*

10. **d.** None of the above
    Reference: *Water Treatment Operator Training Handbook (AWWA)*

11. **b.** nitrite
    Reference: *Water Treatment Operator Training Handbook (AWWA)*

12. **a.** 50 mL, an unpreserved, at the entry point to the distribution system
    Reference: *Water System Operations (WSO) Series (AWWA)*

13. **b.** pH meter, millivolt scale
    Reference: *Water System Operations (WSO) Series (AWWA)*

14. **b.** certain radiological samples
    Reference: *Water System Operations (WSO) Series (AWWA)*

15. **d.** Both a and b
    Reference: *Water System Operations (WSO) Series (AWWA)*

16. **d.** temperature, HAA5, TTHM
    Reference: *Water Treatment Operator Training Handbook (AWWA)*

17. **a.** Provide size and number
Reference: *Water System Operations (WSO) Series (AWWA)*

18. **b.** $NH_4Cl$
Reference: *Water Treatment Operator Training Handbook (AWWA)*

19. **a.** contaminant changes for various reasons
Reference: *Water System Operations (WSO) Series (AWWA)*

20. **c.** fecal coliforms and *E. coli.*
Reference: *Water System Operations (WSO) Series (AWWA)*

21. **c.** organic compounds
Reference: *Water System Operations (WSO) Series (AWWA)*

22. **c.** 9 years
Reference: *Water System Operations (WSO) Series (AWWA)*

23. **d.** no gas has, 48 hours
Reference: *Water System Operations (WSO) Series (AWWA)*

24. **b.** Alkalinity caused by $HCO^{-3}$
Reference: *Water System Operations (WSO) Series (AWWA)*

25. **c.** Langelier saturation index, coupon testing
Reference: *Water System Operations (WSO) Series (AWWA)*

26. **d.** iodometric method
Reference: *Water System Operations (WSO) Series (AWWA)*

27. **b.** 24-hr, 48-hr, confirmed
Reference: *Water System Operations (WSO) Series (AWWA)*

28. **a.** decreases, increases
Reference: *Water System Operations (WSO) Series (AWWA)*

29. **b.** Zeta potential
Reference: *Water System Operations (WSO) Series (AWWA)*

30. **c.** $Fe_2(SO_4)_3 \cdot 3H_2O$
Reference: *Water System Operations (WSO) Series (AWWA)*

31. **b.** yellow, brilliant green bile (BGB)
Reference: *Water System Operations (WSO) Series (AWWA)*

32. **d.** increases, decreases
Reference: *Water System Operations (WSO) Series (AWWA)*

33. **b.** gas, bubbles, 48-hr
Reference: *Water System Operations (WSO) Series (AWWA)*

34. **c.** Soft glass
Reference: *Water System Operations (WSO) Series (AWWA)*

35. **c.** slowly cool, weighing, dust- and moisture-free
Reference: *Water System Operations (WSO) Series (AWWA)*

36. **d.** low alkalinity, saturated with oxygen
Reference: *Water System Operations (WSO) Series (AWWA)*

37. **d.** acid-dichromate
Reference: *Water System Operations (WSO) Series (AWWA)*

38. **c.** Turbidity meters, fixed, qualitative
Reference: *Water System Operations (WSO) Series (AWWA)*

39. **a.** 18–24 hr, lauryl tryptose broth (LTB), 24–48 hr
Reference: *Water System Operations (WSO) Series (AWWA)*

40. **d.** Gooch crucible
Reference: *Water System Operations (WSO) Series (AWWA)*

41. **d.** quarterly until the running annual average is below the MCL
Reference: *Water System Operations (WSO) Series (AWWA)*

42. **c.** Volumetric, Class A
Reference: *Water System Operations (WSO) Series (AWWA)*

43. **c.** Graduated cylinder
Reference: *Water System Operations (WSO) Series (AWWA)*

44. **c.** yellow, salmon pink
Reference: *Water System Operations (WSO) Series (AWWA)*

45. **c.** BGB, an EC broth
Reference: *Water System Operations (WSO) Series (AWWA)*

46. **b.** microbiological samples, reagents, the organisms
Reference: *Water System Operations (WSO) Series (AWWA)*

47. **a.** on a major transmission main
Reference: *Water System Operations (WSO) Series (AWWA)*

48. **c.** Surface water
Reference: *Water System Operations (WSO) Series (AWWA)*

49. **c.** hydroxide, carbonate, and bicarbonate alkalinity
Reference: *Water System Operations (WSO) Series (AWWA)*

50. **c.** Alkalinity, pH, temperature, $CaCO_3$, and total dissolved solids
Reference: *Water System Operations (WSO) Series (AWWA)*

51. **a.** Uranium
Reference: *Water Treatment Operator Training Handbook (AWWA)*

52. **c.** Chemical solutions, samples
Reference: *Water System Operations (WSO) Series (AWWA)*

53. **a.** sodium hexametaphosphate
Reference: *Water System Operations (WSO) Series (AWWA)*

54. **a.** Hydrogen
Reference: *Water System Operations (WSO) Series (AWWA)*

55. **a.** fluoresces under a long-wave UV light
Reference: *Water System Operations (WSO) Series (AWWA)*

56. **c.** mass spectrophotometer, solid-phase microextraction
Reference: *Water System Operations (WSO) Series (AWWA)*

57. **c.** yellow to brown
Reference: *Water System Operations (WSO) Series (AWWA)*

58. **d.** Divided by, dissolved organic carbon
Reference: *Water System Operations (WSO) Series (AWWA)*

59. **a.** Muffle oven
Reference: *Water System Operations (WSO) Series (AWWA)*

60. **b.** 2.1 logs removed
Reference: *Water System Operations (WSO) Series (AWWA)*

First determine % removal.

$$\% \text{ Removal} = \frac{(\text{In} - \text{Out})(100\%)}{\text{In}}$$

$$\% \text{ Removal} = \frac{(150 - 1.2)(100\%)}{150} = 99.2\%$$

Next change 99.2% to decimal form by dividing by 100% = 0.992

Then, calculate the log removal.

$$\text{Log Removal} = (\text{Log}_{10})(-)(100.0 - \%\text{ removed in decimal form})$$

$$\text{Log Removal} = (\text{Log}_{10})(-)(100.0 - 0.992) = (\text{Log}_{10})(-)(99.008)$$

$$(\text{Log}_{10}\ 0.008)(-) = (-2.097)(-) = 2.097\text{, round to 2.1 logs removed}$$

61. **a.** saturation
Reference: *Water System Operations (WSO) Series (AWWA)*

62. **c.** What makes up the pH and how much buffering exists
Reference: *Water System Operations (WSO) Series (AWWA)*

63. **d.** temperature, pH, bicarbonate alkalinity, and total dissolved solids
Reference: *Water System Operations (WSO) Series (AWWA)*

64. **a.** Fe to $Fe^{+2}$
Reference: *Water System Operations (WSO) Series (AWWA)*

65. **d.** microhms, 25°C
Reference: *Water System Operations (WSO) Series (AWWA)*

66. **b.** 6 years
Reference: *Water System Operations (WSO) Series (AWWA)*

67. **c.** coliforms
Reference: *Water System Operations (WSO) Series (AWWA)*

68. **d.** Diluted water samples are placed on a plate-count agar, 48–72 hr
Reference: *Water System Operations (WSO) Series (AWWA)*

69. **b.** A frosted band near the top of the pipette

Reference: *Water System Operations (WSO) Series (AWWA)*

70. **a.** mass and fragmentation pattern

Reference: *Water System Operations (WSO) Series (AWWA)*

71. **d.** Polarographic method

Reference: *Water Treatment Operator Training Handbook (AWWA)*

72. **c.** 3–5 μm, *Cryptosporidium*

Reference: *Water Treatment Operator Training Handbook (AWWA)*

73. **c.** Blue and red

Reference: *Water System Operations (WSO) Series (AWWA)*

74. **c.** calcium

Reference: *Water System Operations (WSO) Series (AWWA)*

75. **b.** red-stained, nonspore-forming, rod-shaped bacteria

Reference: *Water System Operations (WSO) Series (AWWA)*

76. **d.** colorless, pink or red

Reference: *Water System Operations (WSO) Series (AWWA)*

77. **d.** Jar tests

Reference: *Water System Operations (WSO) Series (AWWA)*

78. **c.** purple, yellow, gas

Reference: *Water System Operations (WSO) Series (AWWA)*

79. **b.** fluoride ion, reagent

Reference: *Water System Operations (WSO) Series (AWWA)*

## Equipment Operation and Maintenance

1. **c.** $Fe(OH)_3$

Reference: *Water System Operations (WSO) Series (AWWA)*

2. **c.** kilowatt-hours, watts

Reference: *Water System Operations (WSO) Series (AWWA)*

3. **c.** 16,300,000 lb

Reference: *Math for Water Treatment Operators (AWWA)*

First, find the total surface area of the bottom of the tank.

$\text{Area} = (0.785)(\text{Diameter, ft})^2\left(144\ \text{in.}^2/\text{ft}^2\right)$ Substitute known values:

$\text{Area} = (0.785)(30.0\text{ft})(30.0\text{ft})\left(144\ \text{in.}^2/\text{ft}^2\right)$

$\text{Area} = 101{,}736\ \text{in.}^2$ Then calculate the force in pounds:

$\text{Total force, pounds} = \left(\text{Area, in.}^2\right)(\text{psig})$

$\text{Total force, pounds} = \left(101{,}736\ \text{in.}^2\right)(160\ \text{psig}) = 16{,}277{,}760\ \text{lb, round to } 16{,}300{,}000\ \text{lb}$

4. **c.** 5%
   Reference: *Water System Operations (WSO) Series (AWWA)*
5. **b.** 250°F for 15 min
   Reference: *Pumps and Pumping (ACR)*
6. **d.** higher pH, high temperature, chlorite and chlorate
   Reference: *Water Treatment Operator Training Handbook (AWWA)*
7. **c.** heated to burn away, a brush and detergent
   Reference: *Water System Operations (WSO) Series (AWWA)*
8. **c.** AC, DC, voltage surges
   Reference: *Water System Operations (WSO) Series (AWWA)*
9. **c.** copper
   Reference: *Water System Operations (WSO) Series (AWWA)*
10. **d.** switches, higher interrupting capacity, higher current, switches
   Reference: *Water System Operations (WSO) Series (AWWA)*
11. **a.** 25 times
   Reference: *Water System Operations (WSO) Series (AWWA)*
12. **c.** To prevent the regrowth of bacteria in the distribution system
   Reference: *Water System Operations (WSO) Series (AWWA)*
13. **b.** asbestos.
   Reference: *Pumps and Pumping (ACR)*
14. **b.** pumping too much water
   Reference: *Pumps and Pumping (ACR)*
15. **d.** Either a or c
   Reference: *Pumps and Pumping (ACR)*
16. **c.** Autumn
   Reference: *Water System Operations (WSO) Series (AWWA)*
17. **b.** Centrifugal pump with open impellers
   Reference: *Water System Operations (WSO) Series (AWWA)*
18. **d.** 1, larger
   Reference: *Pumps and Pumping (ACR)*
19. **b.** *Legionella*
   Reference: *Water System Operations (WSO) Series (AWWA)*
20. **d.** produces fewer disinfection byproducts, *Cryptosporidium*
   Reference: *Water Treatment Operator Training Handbook (AWWA)*
21. **b.** higher pressures, higher temperatures
   Reference: *Water System Operations (WSO) Series (AWWA)*
22. **c.** 2.1 V
   Reference: *Water System Operations (WSO) Series (AWWA)*

23. **c.** coated with plastic material
Reference: *Water System Operations (WSO) Series (AWWA)*

24. **d.** First do b then c above.
Reference: *Pumps and Pumping (ACR)*

25. **c.** scale, salts
Reference: *Water System Operations (WSO) Series (AWWA)*

26. **b.** high speeds, relatively slow, short distances
Reference: *Water System Operations (WSO) Series (AWWA)*

27. **b.** Finer and greater density
Reference: *Water System Operations (WSO) Series (AWWA)*

28. **a.** Most pumps are self-priming.
Reference: *Pumps and Pumping (ACR)*

29. **c.** Copper sulfate and lime
Reference: *Water Treatment Operator Training Handbook (AWWA)*

30. **a.** SUS
Reference: *Pumps and Pumping (ACR)*

31. **d.** 150 psi, 350 gpm
Reference: *Pumps and Pumping (ACR)*

32. **d.** Both b and c
Reference: *Pumps and Pumping (ACR)*

33. **a.** $946.89
Reference: *Math for Water Treatment Operators (AWWA)*

Equation:

Cost, month = (hp)(h/d)(Number of days)(0.746 kW/hp)(Cost/kW-hr)(Percent efficiency)

Substitute known values:

Cost, month = (155 Hp) (6.77 h/d) (30 days) (0.746 kW/hp) ($0.048/kW) (84%/100%)

Cost, month = $946.89/month

34. **c.** radial load
Reference: *Pumps and Pumping (ACR)*

35. **b.** 90 and 120
Reference: *Pumps and Pumping (ACR)*

36. **d.** All of the above
Reference: *Pumps and Pumping (ACR)*

37. **b.** At least every 3 months
Reference: *Pumps and Pumping (ACR)*

38. **a.** Solvent
Reference: *Pumps and Pumping (ACR)*

39. **d.** Either b or c

Reference: *Pumps and Pumping (ACR)*

40. **d.** Lime

Reference: *Water System Operations (WSO) Series (AWWA)*

41. **a.** it improves coagulation

Reference: *Water System Operations (WSO) Series (AWWA)*

42. **a.** Two to three times more

Reference: *Water System Operations (WSO) Series (AWWA)*

43. **a.** End suction pump

Reference: *Pumps and Pumping (ACR)*

44. **c.** chlorate and chlorite

Reference: *Water System Operations (WSO) Series (AWWA)*

45. **a.** oxidants, 5–10 years

Reference: *Water System Operations (WSO) Series (AWWA)*

46. **a.** pH, oxidation

Reference: *Water System Operations (WSO) Series (AWWA)*

47. **a.** Volts

Reference: *Water System Operations (WSO) Series (AWWA)*

48. **c.** cooled

Reference: *Pumps and Pumping (ACR)*

49. **a.** 1 in.

Reference: *Water System Operations (WSO) Series (AWWA)*

50. **a.** Rolling pin

Reference: *Pumps and Pumping (ACR)*

51. **c.** reverse flow and air scrubbing

Reference: *Water System Operations (WSO) Series (AWWA)*

52. **a.** 7,400 $ft^3$

Reference: *Math for Water Treatment Operators (AWWA)*

Equation: Volume of cone, $ft^3 = 1/3\,(0.785)\,(\text{Diameter, ft})^2\,(\text{Depth, ft})$

Volume, $ft^3 = 1/3\,(0.785)\,(18.1\text{ft})\,(18.1\text{ft})\,(14.5\text{ft}) = 1,243\text{ft}^3$

Next, find the volume of the cylindrical part of the tank.

Equation: Volume, $ft^3 = (0.785)\,(\text{Diameter, ft})^2\,(\text{Depth, ft})$

Volume, $ft^3 = (0.785)\,(18.1\text{ ft})\,(18.1\text{ ft})\,(24\text{ ft}) = 6,172\text{ ft}^3$

Last, add the two volumes for the answer.

Total volume, $ft^3 = 1,243\text{ ft}^3 + 6,172\text{ ft}^3 = 7,415\text{ ft}^3$, round to $7,400\text{ ft}^3$

53. **b.** Concentric reducer, spool, check valve, gate valve

Reference: *Pumps and Pumping (ACR)*

54. **a.** Stored vertically, on end
Reference: *Pumps and Pumping (ACR)*

55. **d.** heated
Reference: *Pumps and Pumping (ACR)*

## Source Water Characteristics

1. **c.** 5–10 years
Reference: *Water System Operations (WSO) Series (AWWA)*

2. **a.** Deposition of organic matter
Reference: *Water System Operations (WSO) Series (AWWA)*

3. **c.** determining the recharge area of a particular aquifer
Reference: *Water System Operations (WSO) Series (AWWA)*

4. **b.** <1 ntu 95%
Reference: *Water Treatment Operator Training Handbook (AWWA)*

5. **b.** sliding and overturning by its own weight
Reference: *Water System Operations (WSO) Series (AWWA)*

6. **d.** horizontal pressure of water and overturning by its own weight
Reference: *Water System Operations (WSO) Series (AWWA)*

7. **b.** most water: silt layer, lowest pumping costs: gravel
Reference: *Water System Operations (WSO) Series (AWWA)*

8. **c.** 25–50 years
Reference: *Water System Operations (WSO) Series (AWWA)*

9. **c.** ammonia concentrations
Reference: *Water System Operations (WSO) Series (AWWA)*

10. **b.** water or fluid, strata
Reference: *Water System Operations (WSO) Series (AWWA)*

11. **d.** naturally or from wastewater discharge, domestic animals, and wild animals.
Reference: *Water System Operations (WSO) Series (AWWA)*

12. **d.** Humic and fulvic acids
Reference: *Water System Operations (WSO) Series (AWWA)*

13. **d.** Creating an artificial wetland, siltation
Reference: *Water System Operations (WSO) Series (AWWA)*

14. **c.** both bacteria and fungi
Reference: *Water System Operations (WSO) Series (AWWA)*

15. **c.** seasonally, daily, early morning
Reference: *Water System Operations (WSO) Series (AWWA)*

16. **b.** a garlic-like taste
Reference: *Water System Operations (WSO) Series (AWWA)*

17. **d.** Asiatic clam *(Corbicula fluminea)*
    Reference: *Water System Operations (WSO) Series (AWWA)*
18. **c.** *Gallionella*
    Reference: *Water System Operations (WSO) Series (AWWA)*
19. **a.** increases, increases
    Reference: *Water System Operations (WSO) Series (AWWA)*
20. **b.** $MnO_2$, $Mn^{+2}$ and $Mn^{+4}$
    Reference: *Water System Operations (WSO) Series (AWWA)*
21. **d.** Both b and c
    Reference: *Water System Operations (WSO) Series (AWWA)*
22. **b.** oocysts
    Reference: *Water System Operations (WSO) Series (AWWA)*
23. **b.** Weekly
    Reference: *Water System Operations (WSO) Series (AWWA)*
24. **c.** Fluorine
    Reference: *Water System Operations (WSO) Series (AWWA)*
25. **a.** its source water
    Reference: *Water System Operations (WSO) Series (AWWA)*
26. **c.** simple to install, relatively close to, boulders
    Reference: *Water System Operations (WSO) Series (AWWA)*
27. **b.** depth of sunlight penetration
    Reference: *Water System Operations (WSO) Series (AWWA)*
28. **c.** fill material, very deep, scarcity of fill, shallow
    Reference: *Water System Operations (WSO) Series (AWWA)*
29. **d.** It becomes anaerobic.
    Reference: *Water System Operations (WSO) Series (AWWA)*
30. **a.** In spring when the average temperature of the air is higher than the water
    Reference: *Water System Operations (WSO) Series (AWWA)*
31. **c.** Algae problem may worsen
    Reference: *Water System Operations (WSO) Series (AWWA)*
32. **d.** Both a and c
    Reference: *Water Treatment Operator Training Handbook (AWWA)*
33. **b.** 3–15 microns
    Reference: *Water System Operations (WSO) Series (AWWA)*

## Security, Safety, Compliance, and Administrative Procedures

1. **c.** Check all SCADA to make sure all processes are functioning
   Reference: *Water System Operations (WSO) Series (AWWA)*

2. **a.** 0.5

Reference: *Water System Operations (WSO) Series (AWWA)*

3. **a.** Proportional control

Reference: *Water Treatment Operator Training Handbook (AWWA)*

4. **b.** 8

Reference: *Water Treatment Operator Training Handbook (AWWA)*

5. **b.** because of characteristics of the water source, installed full-scale best available technology (BAT)

Reference: *Water System Operations (WSO) Series (AWWA)*

6. **b.** educate the affected public

Reference: *Water System Operations (WSO) Series (AWWA)*

7. **b.** daily

Reference: *Water Treatment Operator Training Handbook (AWWA)*

8. **c.** *Cryptosporidium, E. coli,* and turbidity

Reference: *Water System Operations (WSO) Series (AWWA)*

9. **b.** It adds a taste to the water that some people find objectionable.

Reference: *Water System Operations (WSO) Series (AWWA)*

10. **c.** 50 μg/L to 10 μg/L.

Reference: *Water System Operations (WSO) Series (AWWA)*

11. **c.** More expensive

Reference: *Water System Operations (WSO) Series (AWWA)*

12. **c.** Feedback control

Reference: *Water Treatment Operator Training Handbook (AWWA)*

13. **a.** element, atomic number, neutrons

Reference: *Water System Operations (WSO) Series (AWWA)*

14. **c.** 3 pCi/L

Reference: *Water System Operations (WSO) Series (AWWA)*

15. **b.** Stage 1 Disinfectants and Disinfection Byproducts Rule, 1.0 mg/L

Reference: *Water System Operations (WSO) Series (AWWA)*

16. **c.** 20–50 ppm

Reference: *Water Treatment Operator Training Handbook (AWWA)*

17. **c.** USEPA

Reference: *Water System Operations (WSO) Series (AWWA)*

18. **c.** 2.0-log removal inactivation for *Cryptosporidium.*

Reference: *Water System Operations (WSO) Series (AWWA)*

19. **c.** carbon monoxide

Reference: *Water System Operations (WSO) Series (AWWA)*

20. **c.** the office of the state's drinking water program
Reference: *Water System Operations (WSO) Series (AWWA)*

21. **a.** two consecutive sampling periods
Reference: *Water System Operations (WSO) Series (AWWA)*

22. **a.** Calcium
Reference: *Water System Operations (WSO) Series (AWWA)*

23. **b.** <2.0 mg/L
Reference: *Water Treatment Operator Training Handbook (AWWA)*

24. **b.** *Cryptosporidium*
Reference: *Water Treatment Operator Training Handbook (AWWA)*

25. **d.** *Cryptosporidium,* four
Reference: *Water Treatment Operator Training Handbook (AWWA)*

26. **b.** National Primary Drinking Water Standards, turbidity
Reference: *Water System Operations (WSO) Series (AWWA)*

27. **d.** Both a and b
Reference: *Water System Operations (WSO) Series (AWWA)*

28. **c.** large, monitoring regulations and treatment regulations
Reference: *Water System Operations (WSO) Series (AWWA)*

29. **b.** Split treatment
Reference: *Water System Operations (WSO) Series (AWWA)*

30. **b.** Air mixed in with the shipping material
Reference: *Water System Operations (WSO) Series (AWWA)*

31. **a.** Fecal contamination
Reference: *Water Treatment Operator Training Handbook (AWWA)*

32. **a.** two, 800 lb
Reference: *Water Treatment Operator Training Handbook (AWWA)*

33. **b.** Radiological contaminants
Reference: *Water System Operations (WSO) Series (AWWA)*

34. **c.** 0.7 mg/L
Reference: *Water Treatment Operator Training Handbook (AWWA)*

35. **b.** Causes nerve damage and respiratory problems
Reference: *Water System Operations (WSO) Series (AWWA)*

36. **b.** Feed forward control
Reference: *Water Treatment Operator Training Handbook (AWWA)*

37. **b.** Sonic type
Reference: *Water Treatment Operator Training Handbook (AWWA)*

38. **d.** 4.0-log, viruses
Reference: *Water Treatment Operator Training Handbook (AWWA)*

39. **c.** 1.0 mg/L
Reference: *Water Treatment Operator Training Handbook (AWWA)*

40. **c.** 1.0 ntu
Reference: *Water System Operations (WSO) Series (AWWA)*

41. **b.** 10 μg/L.

42. **a.** Riparian rights, civil
Reference: *Water System Operations (WSO) Series (AWWA)*

43. **b.** Each monitoring location has to comply with the running annual average.
Reference: *Water Treatment Operator Training Handbook (AWWA)*

44. **b.** 10 mg/L, magnesium hardness
Reference: *Water System Operations (WSO) Series (AWWA)*

45. **b.** USEPA
Reference: *Water Treatment Operator Training Handbook (AWWA)*

46. **b.** Total organic carbon
Reference: *Water Treatment Operator Training Handbook (AWWA)*

47. **b.** prior appropriation doctrine
Reference: *Water System Operations (WSO) Series (AWWA)*

48. **d.** There is none.
Reference: *Water System Operations (WSO) Series (AWWA)*

49. **d.** 300
Reference: *Water Treatment Operator Training Handbook (AWWA)*

50. **a.** 3 ppm
Reference: *Water Treatment Operator Training Handbook (AWWA)*

51. **d.** 30 μg/L
Reference: *Water System Operations (WSO) Series (AWWA)*

52. **b.** continuously, daily
Reference: *Water System Operations (WSO) Series (AWWA)*

53. **b.** *Vibrio*
Reference: *Water System Operations (WSO) Series (AWWA)*

54. **d.** Lime sludge
Reference: *Water System Operations (WSO) Series (AWWA)*

55. **b.** 0.2 mg/L
Reference: *Water System Operations (WSO) Series (AWWA)*

56. **b.** It involves the evaluation of groundwater to determine the degree of contamination.
Reference: *Water System Operations (WSO) Series (AWWA)*

57. **a.** Ozone
Reference: *Basic Microbiology for Drinking Water (AWWA)*

58. **b.** pH
Reference: *Water System Operations (WSO) Series (AWWA)*

59. **b.** granular activated carbon
Reference: *Water System Operations (WSO) Series (AWWA)*

60. **a.** chemical release, Risk Management Rule, OSHA
Reference: *Water Treatment Operator Training Handbook (AWWA)*

61. **b.** detection limit, gross alpha, gross beta emitters
Reference: *Water System Operations (WSO) Series (AWWA)*

62. **c.** lifetime distribution system evaluation
Reference: *Water System Operations (WSO) Series (AWWA)*

63. **d.** Anemia
Reference: *Water System Operations (WSO) Series (AWWA)*

64. **d.** All of the above
Reference: *Basic Microbiology for Drinking Water (AWWA)*

65. **c.** Biannually
Reference: *Water Treatment Operator Training Handbook (AWWA)*

66. **b.** 0.1%, 30
Reference: *Water System Operations (WSO) Series (AWWA)*

67. **d.** bin, average
Reference: *Water System Operations (WSO) Series (AWWA)*

68. **d.** 0.5 ntu, 95%
Reference: *Water System Operations (WSO) Series (AWWA)*

69. **b.** nausea and vomiting
Reference: *Water System Operations (WSO) Series (AWWA)*

70. **d.** a calculated percentile level
Reference: *Water System Operations (WSO) Series (AWWA)*

71. **b.** Standard 61
Reference: *Water System Operations (WSO) Series (AWWA)*

72. **a.** Interim Enhanced Surface Water Treatment Rule
Reference: *Water System Operations (WSO) Series (AWWA)*

73. **d.** chemical exposure
Reference: *Water System Operations (WSO) Series (AWWA)*

74. **b.** 3.0 log
Reference: *Water Treatment Operator Training Handbook (AWWA)*

# III. Water Distribution, Introductory Answers

## Distribution System Components

1. **b.** Metal scrappers
   Reference: *Water System Operations (WSO) Series (AWWA)*
2. **b.** Bacterial contamination
   Reference: *Water System Operations (WSO) Series (AWWA)*
3. **a.** Large plug
   Reference: *Water System Operations (WSO) Series (AWWA)*
4. **b.** air-and-vacuum relief valve
   Reference: *Water System Operations (WSO) Series (AWWA)*
5. **d.** All of the above
   Reference: *Water Distribution Operator Training Handbook (AWWA)*
6. **d.** Double check valve
   Reference: *Water System Operations (WSO) Series (AWWA)*
7. **d.** To control the direction of flow
   Reference: *Water System Operations (WSO) Series (AWWA)*
8. **b.** pressure gauges
   Reference: *Water System Operations (WSO) Series (AWWA)*
9. **c.** Near street intersections
   Reference: *Water System Operations (WSO) Series (AWWA)*
10. **d.** cubic feet or gallons
    Reference: *Water System Operations (WSO) Series (AWWA)*
11. **b.** Electrical pulse generator type
    Reference: *Water System Operations (WSO) Series (AWWA)*
12. **b.** System pressure
    Reference: *Water System Operations (WSO) Series (AWWA)*
13. **a.** water quality and reliability.
    Reference: *Water Distribution Operator Training Handbook (AWWA)*

14. **a.** Inspect the first check valve
    Reference: *Water System Operations (WSO) Series (AWWA)*

15. **d.** All of the above
    Reference: *Water Distribution Operator Training Handbook (AWWA)*

16. **d.** All of the above
    Reference: *Water System Operations (WSO) Series (AWWA)*

17. **b.** 25 min
    Reference: *Math for Distribution System Operators (AWWA)*

    First, convert the pipe diameter from inches to feet.

    Number of feet = (8.0 in.)(1 ft/12 in.) = 0.667 ft

    Next, determine the volume in gallons for 145 ft of the 8.0-in. pipeline.

    Equation: Volume, gal = (0.785)(Diameter, ft)$^2$(Length, ft)(7.48 gal/ft$^3$)

    However, three volumes need to be flushed.

    Volume, gal = (0.785)(0.667)(0.667)(145 ft)(7.48 gal/ft$^3$)(3 volumes) = 1,136.35 gal

    Then, find the flushing time.

    $$\text{Flushing time, min} = \frac{1,136.35 \text{ gal}}{45 \text{ gpm}} = 25.25 \text{ min, round to 25 min}$$

18. **b.** a primacy agency
    Reference: *Water System Operations (WSO) Series (AWWA)*

19. **a.** organic compounds, permeation
    Reference: *Water System Operations (WSO) Series (AWWA)*

20. **b.** Boilers
    Reference: *Water System Operations (WSO) Series (AWWA)*

21. **b.** ensure the drain valve is completely closed
    Reference: *Water System Operations (WSO) Series (AWWA)*

22. **a.** packing rings or mechanical seals
    Reference: *Water System Operations (WSO) Series (AWWA)*

23. **d.** do not flood, test cocks
    Reference: *Water System Operations (WSO) Series (AWWA)*

24. **c.** To prevent an illegal connection
    Reference: *Water System Operations (WSO) Series (AWWA)*

25. **d.** All of the above
    Reference: *Water System Operations (WSO) Series (AWWA)*

26. **d.** 97.28% meter accuracy
    Reference: *Math for Distribution System Operators (AWWA)*

    $$\text{Equation : Meter accuracy, \%} = \frac{\text{(Meter reading, gal)(100\%)}}{\text{Actual volume, gal}}$$

    $$\text{Meter accuracy, \%} = \frac{(3,685 \text{ L})(100\%)}{3,788 \text{ L}} = 97.28\% \text{ meter accuracy}$$

27. **b.** Wear rings

Reference: *Water System Operations (WSO) Series (AWWA)*

28. **b.** year

Reference: *Water System Operations (WSO) Series (AWWA)*

29. **a.** horizontal: capacity; vertical: head, power, and efficiency

Reference: *Water Treatment Operator Training Handbook (AWWA)*

30. **c.** dry-barrel hydrants

Reference: *Water System Operations (WSO) Series (AWWA)*

31. **c.** State and local codes

Reference: *Water System Operations (WSO) Series (AWWA)*

32. **b.** Sewer

Reference: *Water System Operations (WSO) Series (AWWA)*

33. **c.** 440/480 V

Reference: *Water System Operations (WSO) Series (AWWA)*

34. **a.** magnetic

Reference: *Water System Operations (WSO) Series (AWWA)*

35. **a.** 20 psi

Reference: *Water System Operations (WSO) Series (AWWA)*

36. **c.** 12.3 $ft^3/s$

Reference: *Math for Distribution System Operators (AWWA)*

$$\text{Equation}: ft^3/s = \frac{(\text{mgd})(1{,}000{,}000\ \text{gal})(1\ ft^3)(1\ \text{day})(1\ \text{min})}{1\ \text{MG}\ (7.48\ \text{gal})(1{,}440\ \text{min})(60\ \text{s})}$$

$$\text{or}\ ft^3/s = \frac{(\text{mgd})(1{,}000{,}000\ \text{gal})(1\ ft^3)(1\ \text{day})}{1\ \text{MG}\ (7.48\ \text{gal})(86{,}400\ \text{s})}$$

$$ft^3/s = \frac{(7.94\ \text{mgd})(1{,}000{,}000\ \text{gal})(1\ ft^3)(1\ \text{day})(1\ \text{min})}{1\ \text{MG}\ (7.48\ \text{gal})(1{,}440\ \text{min})(60\ \text{s})} = 12.3\ ft^3/s$$

37. **d.** ANSI/NSF, 61

Reference: *Water System Operations (WSO) Series (AWWA)*

38. **b.** bare pigs

Reference: *Water System Operations (WSO) Series (AWWA)*

39. **c.** gastrointestinal problems

Reference: *Water System Operations (WSO) Series (AWWA)*

40. **c.** 213,844 gal

Reference: *Math for Distribution System Operators (AWWA)*

Equation: Volume, gal = (Cross-sectional area of the pipe, $ft^2$)(Length, ft)(7.48 gal/$ft^3$)

Therefore, we must first convert miles and inches to feet to use the above equation.

First, convert 1.46 miles to feet for the larger pipe. (5,280 ft/mile)(1.46 miles) = 7,708.8 ft

Then, determine the remainder of the feet for the smaller pipe.

1.93 miles − 1.46 miles = 0.47 mile for smaller pipe.

Convert 0.47 mile to feet. (5,280 ft/mile)(0.47 mile) = 2,481.6 ft

Next, convert 18.0 in. to feet. (18.0 in.)(1 ft/12 in.) = 1.50 ft

Then, determine the cross-sectional area of each pipe size.

Equation: Area, $ft^2$ = (0.785)(Diameter, ft)$^2$(Length, ft)(7.48 gal/$ft^3$)

Volume, gal for 2.0-ft-diameter pipe =

(0.785)(2.00 ft)(2.00 ft)(7,708.8 ft)(7.48 gal/$ft^3$) = 181,058 gal

Volume, gal for 1.5-ft diameter pipe =

(0.785)(1.50 ft)(1.50 ft)(2,481.6 ft)(7.48 gal/$ft^3$) = 32,786 gal

Last, add the two volumes together for the total volume.

Total volume, gal = 181,058 gal + 32,786 gal = 213,844 gal

41. **a.** glands
Reference: *Water System Operations (WSO) Series (AWWA)*

42. **b.** Slide valve
Reference: *Water System Operations (WSO) Series (AWWA)*

43. **b.** Pitot
Reference: *Water System Operations (WSO) Series (AWWA)*

44. **b.** Operate storage facilities to enhance mixing
Reference: *Water Distribution Operator Training Handbook (AWWA)*

45. **d.** differential-pressure
Reference: *Water System Operations (WSO) Series (AWWA)*

46. **b.** Stuffing box
Reference: *Water System Operations (WSO) Series (AWWA)*

47. **d.** a certain range of flows, low, low maintenance, larger pipelines
Reference: *Water System Operations (WSO) Series (AWWA)*

48. **d.** clogged, over-register, worn, under-register
Reference: *Water System Operations (WSO) Series (AWWA)*

49. **a.** iron
Reference: *Water System Operations (WSO) Series (AWWA)*

50. **b.** 1,360,000 gal
Reference: *Math for Distribution System Operators (AWWA)*

First, find the volume in gallons for each of the distribution pipes.

Equation: Volume, gal = (0.785)(Diameter, ft)$^2$(Length, ft)(7.48 gal/$ft^3$)

Volume, pipe "A" gal = (0.785)(2.0 ft)(2.0 ft)(489 ft)(7.48 gal/$ft^3$) = 11,485 gal

Volume, pipe "B" gal = (0.785)(1.5 ft)(1.5 ft)(2,655 ft)(7.48 gal/$ft^3$) = 35,077 gal

Next, determine the volume of the storage tank.

Equation: Volume, gal = $(0.785)(\text{Diameter, ft})^2(\text{Height, ft})(7.48 \text{ gal/ft}^3)$

Volume, gal = $(0.785)(91 \text{ ft})(91 \text{ ft})(27.02 \text{ ft})(7.48 \text{ gal/ft}^3) = 1{,}313{,}831$

Last, add the three volumes together for the total volume.

Total vol., gal = 11,485 gal + 35,077 gal + 1,313,831 gal = 1,360,393 gal

51. **a.** proportional to velocity

Reference: *Water Distribution Operator Training Handbook (AWWA)*

52. **a.** To equalize the pressure on the two sides of the gates

Reference: *Water Distribution Operator Training Handbook (AWWA)*

53. **b.** Bituminous

Reference: *Water System Operations (WSO) Series (AWWA)*

54. **b.** the rate of doing work

Reference: *Water System Operations (WSO) Series (AWWA)*

55. **b.** Air-and-vacuum, high, blowoff, low

Reference: *Water System Operations (WSO) Series (AWWA)*

56. **b.** reinforced concrete pressure pipe

Reference: *Water System Operations (WSO) Series (AWWA)*

57. **c.** arterial-loop system

Reference: *Water System Operations (WSO) Series (AWWA)*

58. **b.** 2.46 fps

Reference: *Math for Distribution System Operators (AWWA)*

First, convert the number of gallons per minute to cubic feet per second.

$$\text{Number of ft}^3\text{/s} = \frac{385 \text{ gpm}}{(7.48 \text{ gal/ft}^3)(60 \text{ s/min})} = 0.858 \text{ ft}^3\text{/s}$$

Next, convert the diameter from inches to feet.

Number of ft = (8.00 in.)(1 ft/12 in.) = 0.667 ft

Equation: Flow, $\text{ft}^3$/s = $(\text{Area, ft}^2)(\text{Velocity, fps})$, where the Area = $(0.785)(\text{Diameter, ft})^2$

$0.858 \text{ ft}^3\text{/s} = (0.785)(0.667 \text{ ft})(0.667 \text{ ft})(\text{Flow, fps})$

Rearrange and solve for the flow in fps.

$$\text{Flow, fps} = \frac{0.858 \text{ ft}^3\text{/s}}{(0.785)(0.667 \text{ ft})(0.667 \text{ ft})} = 2.4568 \text{ fps, round to } \mathbf{2.46 \text{ fps}}$$

59. **a.** surges

Reference: *Water System Operations (WSO) Series (AWWA)*

60. **b.** Frame-mounted

Reference: *Water System Operations (WSO) Series (AWWA)*

61. **a.** proportional

Reference: *Water System Operations (WSO) Series (AWWA)*

62. **d.** Both b and c

Reference: *Water Distribution Operator Training Handbook (AWWA)*

63. **d.** 858,000 lb

First, calculate the tank's area.

Equation : Area, $ft^2$= (0.785) (Diameter, ft)$^2$

Area, $ft^2$= (0.785) (49.5 ft)$^2$= (0.785) (49.5 ft) (49.5 ft) = 1, 923.45 $ft^2$

Next, determine the pressure in pounds per $ft^2$ on the tank's bottom.

Equation : Pressure, lb/$ft^2$= $\left(\text{Water density, lb/ft}^2\right)$ (Height, ft)

Pressure, lb/$ft^2$= $\left(62.4\ \text{lb/ft}^3\right)$ (7.15 ft) = 446.16 lb/$ft^2$

Lastly, find the upward force on the bottom of the tank

Equation : Upward force, lb = $\left(\text{Area, ft}^2\right)\left(\text{Pressure, lb/ft}^2\right)$

Upward force, lb = $\left(1,923.45\ \text{ft}^2\right)\left(446.16\ \text{lb/ft}^2\right)$ = 858, 166 lb, round to 858, 000 lb

64. **a.** Correlative rights

Reference: *Water System Operations (WSO) Series (AWWA)*

65. **d.** All the above

Reference: *Water System Operations (WSO) Series (AWWA)*

66. **c.** long-distance transmission mains

Reference: *Water System Operations (WSO) Series (AWWA)*

67. **a.** Setting renewal priorities

Reference: *Water System Operations (WSO) Series (AWWA)*

68. **a.** fire protection

Reference: *Water System Operations (WSO) Series (AWWA)*

69. **c.** Butterfly valve

Reference: *Water Distribution Operator Training Handbook (AWWA)*

70. **b.** 84 gpm

Reference: *Math for Distribution System Operators (AWWA)*

First, convert the pipe diameter from inches to feet.

Pipe diameter, ft = (10.0 in.)(1 ft/12 in.) = 0.833 ft

Next, determine the volume in gallons for 310 ft of the 10-in. pipeline.

Equation: Volume, gal = (0.785)(Diameter, ft)$^2$(Length, ft)(7.48 gal/$ft^3$)

Volume, gal = (0.785)(0.833 ft)(0.833 ft)(310 ft)(7.48 gal/$ft^3$)

Volume, gal = 1,263 gal

Then, find the amount of gallons that were removed.

Gallons removed $= \dfrac{1,263\ \text{gal}}{15\ \text{min}} = 84.2$ gpm

71. **d.** The anodes will disintegrate quickly.

Reference: *Water Distribution Operator Training Handbook (AWWA)*

72. **b.** 0.734

Reference: *Math for Distribution System Operators (AWWA)*

$$\frac{73.4\%}{100\%} = 0.734$$

73. **b.** 18.3 psi

Reference: *Math for Distribution System Operators (AWWA)*

First, calculate the number of cubic feet of water present.

$$\frac{1,375,000 \text{ gal}}{7.48 \text{ gal/ft}^3} = 183,823.53 \text{ ft}^3$$

Then, the number of $\text{ft}^3$ = (0.785)(Diameter, ft)$^2$(Depth, ft)

Solve for depth:

183,823.53 $\text{ft}^3$ = (0.785)(70 ft)(70 ft)(Depth, ft)

$$\text{Depth, ft} = \frac{183,823.53 \text{ ft}^3}{(0.785)(74.5 \text{ ft})(74.5 \text{ ft})}$$

Depth = 42.19 ft

Next, solve for the number of psi at the bottom of the tank.

$$\text{psi} = \frac{\text{Depth, ft}}{2.31 \text{ ft/psi}} = \frac{42.19 \text{ ft}}{2.31 \text{ ft/psi}} = 18.264\text{psi, round to } 18.3\text{psi}$$

74. **a.** a double cover

Reference: *Water System Operations (WSO) Series (AWWA)*

75. **b.** permeable layer, confined by upper and lower layers, impermeable

Reference: *Water Distribution Operator Training Handbook (AWWA)*

76. **b.** 1.99 fps

Reference: *Math for Distribution System Operators (AWWA)*

First, convert the diameter in inches to feet.

Number of feet = (10.0 in.)(1 ft/12 in.) = 0.833

Next, determine the cross-sectional area of the pipe.

Equation: Area, $\text{ft}^2$ = (0.785)(Diameter, ft)$^2$

Area, $\text{ft}^2$ = (0.785)(0.833 ft)(0.833 ft) = 0.5447 ft

Next, determine the $\text{ft}^3$/s.

Number of $\text{ft}^3$/s = (1,287,000 gal/24 h)(1 $\text{ft}^3$/7.48 gal)(1 h/3,600 s) = 1.0847 $\text{ft}^3$/s

Last, determine the flow in fps.

Equation: Flow, $\text{ft}^3$/s = (Area, $\text{ft}^2$)(Velocity, fps)

Rearrange the formula to solve for velocity.

$$\text{Velocity, fps} = \frac{\text{Flow, ft}^3\text{/s}}{\text{Area, ft}^2}$$

Substitution:

$$\text{Velocity, fps} = \frac{1.0847 \text{ ft}^3\text{/s}}{0.5447 \text{ ft}^2} = 1.9914 \text{ fps, round to } 1.99 \text{ fps}$$

III. Water Distribution, Introductory Answers

77. **b.** Mortar
Reference: *Water System Operations (WSO) Series (AWWA)*

78. **c.** dead ends
Reference: *Water System Operations (WSO) Series (AWWA)*

79. **b.** Maintaining adequate system pressure
Reference: *Water Distribution Operator Training Handbook (AWWA)*

80. **a.** Storage tank
Reference: *Water System Operations (WSO) Series (AWWA)*

81. **b.** rotational speed of a shaft
Reference: *Water System Operations (WSO) Series (AWWA)*

82. **b.** air gap, reduced-pressure zone backflow preventer
Reference: *Water System Operations (WSO) Series (AWWA)*

83. **b.** Ultrasonic system
Reference: *Water System Operations (WSO) Series (AWWA)*

84. **a.** Polyvinyl chloride
Reference: *Water System Operations (WSO) Series (AWWA)*

85. **a.** be repumped to boost water pressure after treatment.
Reference: *Water System Operations (WSO) Series (AWWA)*

86. **d.** All of the above
Reference: *Water System Operations (WSO) Series (AWWA)*

87. **c.** ¾-in.
Reference: *Water System Operations (WSO) Series (AWWA)*

88. **c.** alternately toward the flow, then away in a diagonal direction across the pipe.
Reference: *Water System Operations (WSO) Series (AWWA)*

89. **c.** PVC, concrete pressure, lined ductile iron
Reference: *Water System Operations (WSO) Series (AWWA)*

90. **a.** elevated storage is not available
Reference: *Water System Operations (WSO) Series (AWWA)*

91. **c.** of laminar flow, either side
Reference: *Water Distribution Operator Training Handbook (AWWA)*

92. **b.** Double check valve
Reference: *Water System Operations (WSO) Series (AWWA)*

93. **a.** Transmission mains
Reference: *Water System Operations (WSO) Series (AWWA)*

94. **c.** absorption pit
Reference: *Water System Operations (WSO) Series (AWWA)*

95. **c.** 12.8 hours
Reference: *Math for Distribution System Operators (AWWA)*

$$\text{Equation: Detention time, hours} = \frac{\text{(Tank volume)(24 h/d)}}{\text{Flow, gal/d}}$$

Substitution:

First, convert MG to gallons: (3.25 MG)(1,000,000/1 M) = 3,250,000 gal

And (6.10 mgd)(1,000,000/1 M) = 6,100,000 gal/d

$$\text{Detention time, hours} = \frac{(3,250,000 \text{ gal})(24 \text{ h/d})}{6,100,000 \text{ gal/d}} = 12.787 \text{ hours, round to } 12.8 \text{ hours}$$

96. **d.** All of the above

Reference: *Water System Operations (WSO) Series (AWWA)*

97. **b.** Needs special protection in high-chloride soils

Reference: *Water System Operations (WSO) Series (AWWA)*

98. **b.** Annually

Reference: *Water Distribution Operator Training Handbook (AWWA)*

99. **d.** Graphite

Reference: *Water System Operations (WSO) Series (AWWA)*

100. **b.** The meter will under register or not register anything at low flows.

Reference: *Water Distribution Operator Training Handbook (AWWA)*

101. **c.** 15-in., repaired or replaced

102. **d.** All of the above

Reference: *Water Distribution Operator Training Handbook (AWWA)*

103. **c.** Steel

Reference: *Water System Operations (WSO) Series (AWWA)*

104. **c.** flanged meter couplings.

Reference: *Water Distribution Operator Training Handbook (AWWA)*

105. **c.** Concrete

Reference: *Water System Operations (WSO) Series (AWWA)*

106. **a.** They may not be accurate.

Reference: *Water Distribution Operator Training Handbook (AWWA)*

107. **b.** flexible, reinforced diaphragm

Reference: *Water Distribution Operator Training Handbook (AWWA)*

108. **a.** plug

Reference: *Water System Operations (WSO) Series (AWWA)*

109. **b.** 152 psi

Reference: *Math for Distribution System Operators (AWWA)*

First, subtract the number of feet above the bottom of the lake from the total depth.

Depth at point in question = 367 ft − 15 ft = 352 ft

$$\text{Equation: psi} = \frac{\text{Depth, ft}}{2.31 \text{ ft/psi}}$$

Substitution:

$$\text{psi} = \frac{352 \text{ ft}}{2.31 \text{ ft/psi}} = 152.38 \text{ psi, round to } 152 \text{ psi}$$

110. **b.** 220/240 V

Reference: *Water System Operations (WSO) Series (AWWA)*

111. **b.** Plastic encased pipeline

Reference: *Water System Operations (WSO) Series (AWWA)*

112. **b.** 12,300 $ft^2$

Reference: *Math for Distribution System Operators (AWWA)*

Equation: Area = $\pi(r)^2$, where $\pi$ = 3.14.

Area of tank = (3.14)(62.5 ft)(62.5 ft) = 12,266 ft, round to 12,300 $ft^2$

113. **b.** Pressure-relief valves

Reference: *Water System Operations (WSO) Series (AWWA)*

114. **b.** 3–5 years

Reference: *Water Distribution Operator Training Handbook (AWWA)*

115. **a.** 11,500 gal

Reference: *Water System Operations (WSO) Series (AWWA)*

Give answer to 3 significant figures.

$$\text{Equation: Volume, ft}^3 = \frac{(b_1+b_2)\,(\text{Height, ft})\,(\text{Length, ft})}{2}$$

$$\text{Volume, ft}^3 = \frac{(4.2\ \text{ft}+7.9\ \text{ft})(3.1\ \text{ft})(82.1\ \text{ft})}{2}$$

$$\text{Volume, ft}^3 = \frac{(12.1\text{ft})\,(3.1\text{ft})\,(82.1\text{ft})}{2} = \frac{3{,}079.571}{2} = 1{,}539.79\ \text{ft}^3$$

Next, find the number of gallons.

$$\text{Number of gal} = \left(1{,}539.79\ \text{ft}^3\right)\left(7.48\ \text{gal/ft}^3\right) = 11{,}517.6\ \text{gal, round to } 11{,}500\ \text{gal}$$

116. **d.** All of the above

Reference: *Water Distribution Operator Training Handbook (AWWA)*

117. **a.** start and stop flows

Reference: *Water System Operations (WSO) Series (AWWA)*

118. **c.** 0.315 $m^3/s$

Reference: *Math for Distribution System Operators (AWWA)*

First, convert gallons to liters.

Equation: Number of Liters = (Number of gpm)(1 min/60 s)(3.785 L/gal)

Number of Liters = (4,994 gpm)(1 min/60 s)(3.785 L/gal) = 315.038 L/s

Next convert liters per second (L/s) to cubic meters per second.

$$\text{Equation : Number m}^3/\text{s} = \frac{\text{Flow, L/s}}{1{,}000\text{L/m}^3}$$

$$\text{Number m}^3/\text{s} = \frac{315.038\ \text{L/s}}{1{,}000\ \text{L/m}^3} = 0.315\ \text{m}^3/\text{s}$$

119. **c.** 31 hours 32 min

Reference: *Math for Distribution System Operators (AWWA)*

First, determine the capacity of the swimming pool in gallons.

Equation: Volume, gal = (Length, ft)(Width, ft)(Depth, ft)(7.48 gal/ft$^3$)\

Volume, gal = (31 ft)(21 ft)(5.4 ft)(7.48 gal/ft$^3$) = 26,295 gal

Next, find the time to fill the pool.

$$\text{Equation : Time, min} = \frac{\text{Volume, gal}}{\text{Rate, gpm}}$$

$$\text{Time, min} = \frac{26,295\text{ gal}}{13.9\text{ gpm}} = 1,891.73\text{ min}$$

Convert minutes to hours.

$$\text{Time, hours} = \frac{1,891.73\text{ min}}{60\text{ min/h}} = 31.53\text{ hours}$$

Next, convert 0.53 hours by multiplying by 60 min/h:

Number of minutes = (0.53 hours)(60 min/h) = 31.8 min, round to 32 min

Thus, time to fill the pool is 31 hours and 32 min

120. **a.** 1.83 ft/s in 10.0-in. pipe

Reference: *Water System Operations (WSO) Series (AWWA)*

Flow in 8.0-in. pipe equals the flow in the 10.0-in. pipe as the flow must remain constant:

$Q_1 = Q_2$

Since Q, flow = (Area)(Velocity), it follows that: (Area 1)(Velocity 1) = (Area 2)(Velocity 2)

First, find the diameters in feet for the 8.0-in. and 10-in. pipes:

Diameter for 8.0-in., ft = 8.0-in(1 ft/12-in.) = 0.667 ft

Diameter for 10.0-in., ft = 10.0-in(1 ft/12-in.) = 0.833 ft

Then determine the areas of each size pipe: Area = (0.785)(Diameter, ft)$^2$

Area 1 (8.0-in.), ft$^2$ = (0.785)(0.667 ft)(0.667 ft) = 0.349 ft$^2$

Area 2 (10.0-in.), ft$^2$ = (0.785)(0.833 ft)(0.833 ft) = 0.545 ft$^2$

Lastly, substitute areas calculated and known velocity in 8.0-in. pipe.

(0.349 ft$^2$)(2.85 ft/s) = (0.545 ft$^2$)(x, ft/s) Solve for x:

$$\text{x, ft/s} = \frac{(0.349\text{ ft}^2)(2.85\text{ ft/s})}{(0.545\text{ ft}^2)} = 1.825\text{ ft/s, round to 1.83 ft/s in 10.0-in. pipe}$$

III. Water Distribution, Introductory Answers

121. **c.** 11,531.65 $ft^2$

Reference: *Math for Distribution System Operators (AWWA)*

First, find the surface area of the tank's wall. Imagine cutting the wall and rolling it out so that it is flat. The length of this now flat wall is simply the circumference of the tank and is defined as (Diameter)(π).

The area of the tank wall then is the length of the wall times the height:

Wall area, $ft^2$ = (Diameter, ft)(π)(Height, ft)

Substitution:

Wall area, $ft^2$ = (65.0 ft)(3.14)(24.0 ft) = 4,898.4 $ft^2$

Next, find the top and bottom surface area.

Top and bottom area, $ft^2$ = (2, top and bottom)(0.785)(Diameter, ft)$^2$

Top and bottom area, $ft^2$ = (2)(0.785)(65.0 ft)(65.0 ft) = 6,633.25 $ft^2$

Total area of tank, $ft^2$ = 4,898.4 $ft^2$ + 6633.25 $ft^2$ = 11,531.65 $ft^2$

122. **a.** Reinforced concrete noncylinder pipe

Reference: *Water System Operations (WSO) Series (AWWA)*

123. **a.** The total feet of head against which a pump must work

Reference: *Water System Operations (WSO) Series (AWWA)*

124. **c.** Flexural strength

Reference: *Water System Operations (WSO) Series (AWWA)*

125. **a.** state and local health authorities

Reference: *Water System Operations (WSO) Series (AWWA)*

126. **c.** in direct proportion, rotor

Reference: *Water System Operations (WSO) Series (AWWA)*

127. **c.** 7.3 mgd

Reference: *Math for Distribution System Operators (AWWA)*

$$\text{Equation : Avg. mgd flow} = \frac{\text{Sum of mgd used each day}}{\text{Total time. davs}}$$

$$\text{Avg. mgd flow} = \frac{7.9 + 8.0 + 6.5 + 6.8 + 7.0 + 7.3 + 7.8}{7 \text{ days}} = 7.33 \text{ mgd, round to } 7.3 \text{ mgd}$$

128. **c.** the vertical distance from point of pressure measurement to the hydraulic grade line

Reference: *Water System Operations (WSO) Series (AWWA)*

129. **c.** industrial, domestic

Reference: *Water System Operations (WSO) Series (AWWA)*

130. **b.** 103 ft

Reference: *Math for Distribution System Operators (AWWA)*

Equation: Circumference = (π)(Diameter) Rearrange the equation to solve for the diameter.

$$\text{Diameter} = \frac{\text{Circumference}}{(\pi)}$$

$$\text{Diameter} = \frac{323 \text{ ft}}{3.14} = 103 \text{ ft}$$

131. **b.** high-carbon spring-steel blades

Reference: *Water System Operations (WSO) Series (AWWA)*

132. **a.** Concrete

Reference: *Water System Operations (WSO) Series (AWWA)*

133. **a.** 10%

Reference: *Water System Operations (WSO) Series (AWWA)*

134. **b.** 13.7 ft

Reference: *Math for Distribution System Operators (AWWA)*

Write the equation, arranging it to solve for the unknown, drawdown.

$$\text{Equation : Drawdown, ft} = \frac{\text{Well yield, gpm}}{\text{Specific yield, gpm/ft}}$$

$$\text{Drawdown, ft} = \frac{129\text{ gpm}}{45\text{ gpm/ft}} = 13.7\text{ ft}$$

135. **c.** Pump station piping

Reference: *Water System Operations (WSO) Series (AWWA)*

136. **a.** SCADA

Reference: *Water System Operations (WSO) Series (AWWA)*

137. **a.** Thrust anchors

Reference: *Water System Operations (WSO) Series (AWWA)*

138. **b.** 2 ft

Reference: *Water System Operations (WSO) Series (AWWA)*

139. **d.** 1 full turn

Reference: *Water System Operations (WSO) Series (AWWA)*

140. **b.** DIP

Reference: *Water System Operations (WSO) Series (AWWA)*

141. **b.** 63,400 gal pumped

Reference: *Math for Distribution System Operators (AWWA)*

Let x = the number of gallons pumped in 24 hours.

$$\frac{\text{x, gal pumped}}{7\text{ hours}} = \frac{435,000\text{ gal}}{48\text{ hours}}$$

$$\text{x, gal pumped} = \frac{(435,000\text{ gal})(7\text{ hours})}{48\text{ hours}} = 63,437.5\text{ gal, round to }63,400\text{ gal pumped}$$

142. **d.** A check valve then a gate valve

Reference: *Water System Operations (WSO) Series (AWWA)*

143. **b.** 2

Reference: *Water System Operations (WSO) Series (AWWA)*

144. **d.** do not wear or damage shaft sleeves

Reference: *Water System Operations (WSO) Series (AWWA)*

145. **a.** globe valve, spring

Reference: *Water Distribution Operator Training Handbook (AWWA)*

146. **c.** 0.058 psi

Reference: *Water System Operations (WSO) Series (AWWA)*

First, find the head loss in feet.

$$\text{Equation : Headloss, ft} = \frac{(\text{velocity})^2}{2(g)}; \text{where } g \text{ is the acceleration of gravity}$$

$$\text{Head loss, ft} = \frac{(2.95\text{ fps})^2}{2\,(32.2\text{ ft/sec}^2)} = \frac{(2.95\text{ fps})\,(2.95\text{ fps})}{64.4\text{ ft/sec}^2} = 0.135\text{ ft}$$

Next, convert head loss in feet to head loss in psi.

Head loss, psi = (0.135 ft) (0.433 psi/ft) = 0.058 psi

147. **d.** push-on joint

Reference: *Water System Operations (WSO) Series (AWWA)*

148. **c.** Backsiphonage

Reference: *Water System Operations (WSO) Series (AWWA)*

149. **b.** Rotary pump

Reference: *Water System Operations (WSO) Series (AWWA)*

150. **b.** Three

Reference: *Water System Operations (WSO) Series (AWWA)*

151. **d.** All of the above

Reference: *Water System Operations (WSO) Series (AWWA)*

152. **a.** AWWA, National Science Foundation (NSF)

Reference: *Water System Operations (WSO) Series (AWWA)*

153. **b.** 27.1 hours

Reference: *Math for Distribution System Operators (AWWA)*

First, convert gallons per minute to gallons per hour (gph): (1,750 gpm)(60 min/h) = 105,000 gph

Next, convert MG to gallons.

Number of gal = (2.85 MG)(1,000,000/1 M) = 2,850,000 gal

Equation: Time, hours = Number of gallons/gph

Time, hours = 2,850,000 gal/105,000 gph = 27.143 hours, round to 27.1 hours

154. **a.** low friction loss, rotor

Reference: *Water System Operations (WSO) Series (AWWA)*

155. **d.** Centrifugal pump

Reference: *Water System Operations (WSO) Series (AWWA)*

156. **a.** Submersible pump

Reference: *Water System Operations (WSO) Series (AWWA)*

157. **d.** 510 $yd^3$

Reference: *Math for Distribution System Operators (AWWA)*

First, determine the number of cubic feet excavated from the trench.

Equation: Volume, $ft^3$ = (Length, ft)(Width, ft)(Depth, ft)

Volume, $ft^3$ = (728 ft)(4.2 ft)(4.5 ft) = 13,759.2 $ft^3$, round to 13,800 $ft^3$

Next, find the number of cubic yards.

$$\text{Number of yd}^3 = \frac{13,759.2\ \text{ft}^3}{27\ \text{ft}^3/\text{yd}^3} = 509.6\ \text{yd}^3\text{, round to 510 yd}^3$$

158. **c.** annually

Reference: *Water System Operations (WSO) Series (AWWA)*

159. **d.** 609,000 lb

Reference: *Math for Distribution System Operators (AWWA)*

First, calculate the tank's area.

Equation: Area, ft = (Length, ft)(Width, ft)

Area, $ft^2$ = (50.0 ft)(24.0 ft) = 1,200 $ft^2$

Next, determine the pressure in pounds per square feet on the tank's bottom.

Equation: Pressure, $lb/ft^2$ = (Water Density, $lb/ft^3$)(Height, ft)

Pressure, $lb/ft^2$ = (62.4 $lb/ft^3$)(8.13 ft) = 507.312 $lb/ft^2$

Last, find the upward force on the bottom of the tank.

Equation: Upward force, lb = (Area, $ft^2$)(Pressure, $lb/ft^2$)

Upward force, lb = (1,200 $ft^2$)(507.312 $lb/ft^2$) = 608,774.4 lb, round to 609,000 lb

160. **a.** 26.5 hr

Reference: *Math for Distribution System Operators (AWWA)*

First, convert gpm to gallons per hour (gph): (1,100 gpm)(60 min/h) = 66,000 gph

Next, convert 1.75 MG to gallons:

(1.75 MG)(1,000,000 gals/1 M) = 1,750,000 gal

Equation: Time, hours = Number of gallons/gph

Time, hours = 1,750,000 gal/66,000 gph = 26.5 hours

161. **d.** autumn

Reference: *Water System Operations (WSO) Series (AWWA)*

162. **b.** lubrication style, 6 months or at annual intervals

Reference: *Water System Operations (WSO) Series (AWWA)*

163. **a.** 60 cycles or 60 Hz

Reference: *Water System Operations (WSO) Series (AWWA)*

164. **c.** Too high a space requirement
Reference: *Water System Operations (WSO) Series (AWWA)*

165. **d.** All of the above
Reference: *Water Distribution Operator Training Handbook (AWWA)*

166. **a.** other than a service line
Reference: *Water System Operations (WSO) Series (AWWA)*

167. **a.** Four
Reference: *Water System Operations (WSO) Series (AWWA)*

168. **a.** 20 psi
Reference: *Water System Operations (WSO) Series (AWWA)*

169. **b.** 81.4 psi
Reference: *Math for Distribution System Operators (AWWA)*

Equation: Pressure head, ft = (Pressure, psi)(2.31 ft/psi)

Rearrange to solve for pressure in psi:

$$\text{Pressure, psi} = \frac{\text{Pressure head, ft}}{2.31\ \text{ft/psi}}$$

Substitute known values:

$$\text{Pressure, psi} = \frac{188\ \text{ft}}{2.31\ \text{ft/psi}} = 81.4\ \text{psi}$$

170. **d.** 7,790 $ft^2$
Reference: *Math for Distribution System Operators (AWWA)*

Equation: Area = $(\pi)r^2$

Area = (3.14)(49.8 ft)(49.8 ft) = 7,787 $ft^2$, round to 7,790 $ft^2$

171. **d.** All of the above
Reference: *Water System Operations (WSO) Series (AWWA)*

172. **c.** Vertical turbine pump
Reference: *Water System Operations (WSO) Series (AWWA)*

173. **a.** appropriative rights
Reference: *Water System Operations (WSO) Series (AWWA)*

174. **c.** artesian well
Reference: *Water Distribution Operator Training Handbook (AWWA)*

175. **b.** compound meter
Reference: *Water System Operations (WSO) Series (AWWA)*

176. **d.** 132 min
Reference: *Math for Distribution System Operators (AWWA)*

First, convert the pipe diameter from inches to feet.

Number of feet = (10.0 in.)(1 ft/12 in.) = 0.833 ft

Next, determine the volume in gallons for 1,625 ft of the 10-in. pipeline.

Equation: Volume, gal = (0.785)(Diameter, ft)$^2$(Length, ft)(7.48 gal/ft$^3$)

Volume, gal = (0.785)(0.833 ft)(0.833 ft)(1,625 ft)(7.48 gal/ft$^3$) = 6,620.86 gal

Then, find the flushing time.

$$\text{Flushing time, min} = \frac{6,620.86 \text{ gal}}{50.0 \text{ gpm}} = 132.4 \text{ min, round to } 132 \text{ min}$$

177. **b.** equal to or greater than that

Reference: *Water System Operations (WSO) Series (AWWA)*

178. **a.** Turbine pump

Reference: *Water Distribution Operator Training Handbook (AWWA)*

179. **a.** Cast-iron pipe

Reference: *Water System Operations (WSO) Series (AWWA)*

180. **b.** 2.01 ft/s

Reference: *Math for Distribution System Operators (AWWA)*

First, convert the diameter from inches to feet. Number of feet = (8.0 in.)(1 ft/12 in.) = 0.667 ft

Next, convert the number of gallons per minute to cubic feet per second.

$$\text{Number of ft}^3/\text{s} = \frac{315 \text{ gpm}}{(7.48 \text{ gal/ft}^3)(60 \text{ s/min})} = 0.702 \text{ ft}^3/\text{s}$$

Equation for flow is Flow, ft$^3$/s = (Area, ft$^2$)(Velocity, fps);

Where the area = (0.785)(Diameter, ft)$^2$

0.702 ft$^3$/s = (0.785)(0.667 ft)(0.667 ft)(Flow, fps)

Rearrange the problem to solve for flow:

$$\text{Flow, ft/s} = \frac{0.702 \text{ ft}^3/\text{s}}{(0.785)(0.667 \text{ ft})(0.667 \text{ ft})} = 2.010 \text{ ft/s}$$

181. **d.** an altitude valve, unrestricted

Reference: *Water System Operations (WSO) Series (AWWA)*

182. **b.** Ductile iron

Reference: *Water System Operations (WSO) Series (AWWA)*

183. **a.** blue baby syndrome (methemoglobinemia)

Reference: *Water System Operations (WSO) Series (AWWA)*

184. **c.** supply, downstream, relief valve's spring tension

Reference: *Water System Operations (WSO) Series (AWWA)*

## Equipment Installation, Operation, and Maintenance

1. **a.** corporation stop

Reference: *Water System Operations (WSO) Series (AWWA)*

2. **c.** 6 hr

   Reference: *Water Distribution Operator Training Handbook (AWWA)*

3. **b.** 1–2 ft

   Reference: *Water System Operations (WSO) Series (AWWA)*

4. **b.** The local fire department

   Reference: *Water System Operations (WSO) Series (AWWA)*

5. **c.** 4.76 gpm/ft

   Reference: *Water System Operations (WSO) Series (AWWA)*

   Equation: $\text{Specific yield, gpm/ft} = \dfrac{\text{Well yield, gpm}}{\text{Drawdown, ft}}$

   $\text{Specific yield, gpm/ft} = \dfrac{107 \text{ gpm}}{22.5 \text{ ft}} = 4.755 \text{ gpm/ft, round to } 4.76 \text{ gpm/ft}$

6. **d.** Plug and ball

   Reference: *Water System Operations (WSO) Series (AWWA)*

7. **d.** All of the above

   Reference: *Water System Operations (WSO) Series (AWWA)*

8. **a.** Dry-barrel hydrants

   Reference: *Water System Operations (WSO) Series (AWWA)*

9. **a.** tensile and flexural strength

   Reference: *Water System Operations (WSO) Series (AWWA)*

10. **b.** 15%

    Reference: *Water System Operations (WSO) Series (AWWA)*

11. **c.** bronze, machined, hydrant or valve

    Reference: *Water System Operations (WSO) Series (AWWA)*

12. **b.** 5-psi

    Reference: *Water System Operations (WSO) Series (AWWA)*

13. **c.** five

    Reference: *Water System Operations (WSO) Series (AWWA)*

14. **d.** All of the above

    Reference: *Water System Operations (WSO) Series (AWWA)*

15. **b.** static level, drawdown, pump is losing efficiency

    Reference: *Water Distribution Operator Training Handbook (AWWA)*

16. **b.** 2.5-in., 4.5-in., pumper suction hose

    Reference: *Water System Operations (WSO) Series (AWWA)*

17. **b.** crystallizes

    Reference: *Water System Operations (WSO) Series (AWWA)*

18. **c.** solder joints, flare and/or compression joints

    Reference: *Water Distribution Operator Training Handbook (AWWA)*

19. **c.** Flapper valve
Reference: *Water System Operations (WSO) Series (AWWA)*

20. **a.** distribute surface loads
Reference: *Water Distribution Operator Training Handbook (AWWA)*

21. **b.** Warm-climate hydrants
Reference: *Water System Operations (WSO) Series (AWWA)*

22. **a.** ball valves
Reference: *Water Distribution Operator Training Handbook (AWWA)*

23. **b.** Fewer customer complaints
Reference: *Water Distribution Operator Training Handbook (AWWA)*

24. **d.** well-graded, 1-in.
Reference: *Water System Operations (WSO) Series (AWWA)*

25. **c.** copper, steel, asbestos cement, concrete
Reference: *Water System Operations (WSO) Series (AWWA)*

26. **a.** Sandy soils
Reference: *Water System Operations (WSO) Series (AWWA)*

27. **c.** concrete blocks, inside
Reference: *Water Distribution Operator Training Handbook (AWWA)*

28. **c.** shear, tensile, flexural
Reference: *Water System Operations (WSO) Series (AWWA)*

29. **d.** All of the above
Reference: *Water System Operations (WSO) Series (AWWA)*

30. **c.** calcium hypochlorite
Reference: *Water Distribution Operator Training Handbook (AWWA)*

31. **b.** electrical resistivity
Reference: *Water System Operations (WSO) Series (AWWA)*

32. **d.** Bypass valve
Reference: *Water System Operations (WSO) Series (AWWA)*

33. **c.** It eliminates entry into the building.
Reference: *Water System Operations (WSO) Series (AWWA)*

34. **b.** upper, dense
Reference: *Water System Operations (WSO) Series (AWWA)*

35. **c.** Double check valve
Reference: *Water System Operations (WSO) Series (AWWA)*

36. **c.** Flow-through systems
Reference: *Water Distribution Operator Training Handbook (AWWA)*

37. **c.** 300 mg/L, 3 hr
Reference: *Water System Operations (WSO) Series (AWWA)*

38. **c.** Centrifugal-jet pump

Reference: *Water System Operations (WSO) Series (AWWA)*

39. **c.** 76 psi

Reference: *Math for Distribution System Operators (AWWA)*

First, calculate the column of water in feet.

Number of feet = Total depth of well – Depth to water – Number of feet above bottom

Number of ft = 306.1 – 124.7 ft – 5.8 ft = 175.6 ft

Next, find the pressure in psi. This time, use 0.433 psi/ft.

Equation: Pressure, psi = (Number of ft)(0.433 psi/ft)

Pressure, psi = (175.6 ft)(0.433 psi/ft) = 76.03 psi, round to 76 psi

40. **d.** 92.2 ft

Reference: *Math for Distribution System Operators (AWWA)*

Equation: Drawdown, ft = Pumping water level, ft – Static water level, ft

Rearrange the equation to solve for static water level.

Static water level, ft = Pumping water level, ft – Drawdown, ft

Substitute known values:

Static water level, ft = 106.4 ft – 14.2 ft = 92.2 ft

41. **d.** All of the above

Reference: *Water Distribution Operator Training Handbook (AWWA)*

42. **b.** Pumping

Reference: *Water System Operations (WSO) Series (AWWA)*

43. **b.** 110 $yd^3$

Reference: *Math for Distribution System Operators (AWWA)*

First, find the number of cubic feet. Equation: Volume, $ft^3$ = (Length, ft)(Width, ft)(Depth, ft)

Volume, $ft^3$ = (112 ft)(4.5 ft)(5.9 ft) = 2,973.6 $ft^3$

Next, convert cubic feet to cubic yards. Know 1 $yd^3$ is 27 $ft^3$, thus:

$$\frac{2,973.6\,ft^3}{27\,ft^3/yd^3} = 110.13\,yd^3\text{, round to }110\,yd^3$$

44. **a.** nozzle

Reference: *Water System Operations (WSO) Series (AWWA)*

45. **a.** 2–4 in.

Reference: *Water Distribution Operator Training Handbook (AWWA)*

46. **b.** Fuses or circuit breakers

Reference: *Water System Operations (WSO) Series (AWWA)*

47. **a.** 2-in.

Reference: *Water System Operations (WSO) Series (AWWA)*

48. **a.** pipe friction losses
Reference: *Water System Operations (WSO) Series (AWWA)*

49. **d.** the same material as water main materials
Reference: *Water Distribution Operator Training Handbook (AWWA)*

50. **c.** expansive under the action of freezing but will not fracture
Reference: *Water Distribution Operator Training Handbook (AWWA)*

51. **b.** Clay soils
Reference: *Water System Operations (WSO) Series (AWWA)*

52. **d.** All of the above
Reference: *Water System Operations (WSO) Series (AWWA)*

53. **a.** 1 min, 10-min
Reference: *Water System Operations (WSO) Series (AWWA)*

54. **c.** compacted soil so the barrel of the pipe has continuous firm support
Reference: *Water System Operations (WSO) Series (AWWA)*

55. **d.** 8.77 gpm/ft
Reference: *Math for Distribution System Operators (AWWA)*

$$\text{Equation : Specific yield, gpm/ft} = \frac{\text{Well yield, gpm}}{\text{Drawdown, ft}}$$

$$\text{Specific yield, gpm/ft} = \frac{121\text{ gpm}}{13.8\text{ ft}} = 8.768\text{ gpm/ft, round to } 8.77\text{ gpm/ft}$$

56. **a.** low-flow, low-pressure
Reference: *Water System Operations (WSO) Series (AWWA)*

57. **c.** 25 mg/L
Reference: *Water System Operations (WSO) Series (AWWA)*

58. **d.** Both b and c
Reference: *Water Distribution Operator Training Handbook (AWWA)*

59. **c.** Gray cast-iron pipe
Reference: *Water System Operations (WSO) Series (AWWA)*

60. **a.** coupon, saved for checking its condition
Reference: *Water System Operations (WSO) Series (AWWA)*

61. **a.** granular material with few fines is being used
Reference: *Water System Operations (WSO) Series (AWWA)*

62. **d.** On the shaft, in or near the bearings
Reference: *Water System Operations (WSO) Series (AWWA)*

63. **c.** Wet-barrel hydrants
Reference: *Water System Operations (WSO) Series (AWWA)*

64. **b.** ductile-iron pipe
Reference: *Water System Operations (WSO) Series (AWWA)*

65. **b.** Chemical metering pump
Reference: *Water Distribution Operator Training Handbook (AWWA)*

66. **a.** a small leak
Reference: *Water System Operations (WSO) Series (AWWA)*

67. **d.** All of the above
Reference: *Water Distribution Operator Training Handbook (AWWA)*

68. **a.** 6%
Reference: *Water System Operations (WSO) Series (AWWA)*

69. **a.** Horizontal gate valve
Reference: *Water System Operations (WSO) Series (AWWA)*

70. **d.** 2.5 ft/s, 3.5 ft/s
Reference: *Water System Operations (WSO) Series (AWWA)*

71. **b.** Tapping valve
Reference: *Water System Operations (WSO) Series (AWWA)*

72. **a.** 150 psi
Reference: *Water System Operations (WSO) Series (AWWA)*

73. **c.** plugged with the standard type of pipe material that is being used
Reference: *Water System Operations (WSO) Series (AWWA)*

74. **d.** Corporation stops
Reference: *Water System Operations (WSO) Series (AWWA)*

75. **d.** Both b and c
Reference: *Water Distribution Operator Training Handbook (AWWA)*

76. **b.** large-diameter mains
Reference: *Water System Operations (WSO) Series (AWWA)*

77. **c.** ANSI/AWWA Standard C652
Reference: *Water Distribution Operator Training Handbook (AWWA)*

78. **b.** thoroughly clean the exterior of the pipe
Reference: *Water Distribution Operator Training Handbook (AWWA)*

79. **b.** Steel pipe, essentially all
Reference: *Water System Operations (WSO) Series (AWWA)*

80. **b.** Thermal-overload relays
Reference: *Water System Operations (WSO) Series (AWWA)*

81. **a.** 2 ft.
Reference: *Water System Operations (WSO) Series (AWWA)*

82. **b.** a flanged connection, advantage, cannot
Reference: *Water System Operations (WSO) Series (AWWA)*

83. **a.** whenever reasonably possible
Reference: *Water System Operations (WSO) Series (AWWA)*

84. **b.** A calibrated makeup reservoir
Reference: *Water System Operations (WSO) Series (AWWA)*

85. **d.** Electrical line
Reference: *Water System Operations (WSO) Series (AWWA)*

86. **c.** clays
Reference: *Water System Operations (WSO) Series (AWWA)*

87. **b.** copper tubing, iron pipe thread, plastic pipe
Reference: *Water System Operations (WSO) Series (AWWA)*

88. **b.** Gate valve
Reference: *Water System Operations (WSO) Series (AWWA)*

89. **b.** 1 month
Reference: *Water System Operations (WSO) Series (AWWA)*

90. **b.** is all the way to the face of the bell, spigot
Reference: *Water Distribution Operator Training Handbook (AWWA)*

91. **a.** no greater residual than the feedwater
Reference: *Water Distribution Operator Training Handbook (AWWA)*

92. **a.** drawdown remains the same but the static level drops
Reference: *Water Distribution Operator Training Handbook (AWWA)*

93. **c.** Mueller, larger, iron-pipe
Reference: *Water System Operations (WSO) Series (AWWA)*

94. **b.** monthly, annually
Reference: *Water System Operations (WSO) Series (AWWA)*

95. **b.** partially or totally shut the line down
Reference: *Water System Operations (WSO) Series (AWWA)*

96. **b.** 6 in., pressure and flow, at the end
Reference: *Water Distribution Operator Training Handbook (AWWA)*

97. **a.** The water should be tested again.
Reference: *Water System Operations (WSO) Series (AWWA)*

98. **d.** hydrometer, antifreeze and/or water, alkaline water
Reference: *Water System Operations (WSO) Series (AWWA)*

99. **d.** Durability
Reference: *Water System Operations (WSO) Series (AWWA)*

100. **a.** partially backfilled
Reference: *Water System Operations (WSO) Series (AWWA)*

101. **a.** diesel, fuel-injection
Reference: *Water System Operations (WSO) Series (AWWA)*

102. **a.** wrinkle bending
Reference: *Water System Operations (WSO) Series (AWWA)*

103. **a.** the regulatory requirements
Reference: *Water Distribution Operator Training Handbook (AWWA)*

104. **d.** All of the above
Reference: *Water Distribution Operator Training Handbook (AWWA)*

105. **a.** 6–12 months, environmental conditions
Reference: *Water System Operations (WSO) Series (AWWA)*

106. **c.** a leveling board
Reference: *Water System Operations (WSO) Series (AWWA)*

107. **b.** diffuser
Reference: *Water System Operations (WSO) Series (AWWA)*

108. **b.** well face or screen, bailer
Reference: *Water System Operations (WSO) Series (AWWA)*

109. **b.** Not clean
Reference: *Water System Operations (WSO) Series (AWWA)*

110. **c.** Restricted flow
Reference: *Water System Operations (WSO) Series (AWWA)*

111. **b.** 5 days
Reference: *Water System Operations (WSO) Series (AWWA)*

112. **c.** ice, flap valves, pressure, and vacuum
Reference: *Water System Operations (WSO) Series (AWWA)*

113. **c.** closed, vertical
Reference: *Water System Operations (WSO) Series (AWWA)*

114. **d.** 1 year
Reference: *Water System Operations (WSO) Series (AWWA)*

115. **c.** amplifiers
Reference: *Water System Operations (WSO) Series (AWWA)*

116. **b.** New, should not, closed
Reference: *Water System Operations (WSO) Series (AWWA)*

## Disinfection Monitoring, Evaluation, Adjustment, and Laboratory Analysis/Interpretation

1. **b.** 35.3 lb of $Ca(OCl)_2$

First, calculate the number of gallons in the pipeline.

Equation: Pipe volume, gal = (0.785)(Diameter, ft)$^2$(Length, ft)(7.48 gal/ft$^3$)

Pipe volume, gal = (0.785)(3.00 ft)(3.00 ft)(1,040 ft)(7.48 gal/ft$^3$) = 54,960 gal

Convert gallons to million gallons.

Pipe volume, MG = (54,960 gal)(1 M/1,000,000) = 0.05496 MG

Next, determine the number of pounds of calcium hypochlorite [$Ca(OCl)_2$] needed using the modified "pounds" formula.

$$Ca(OCl)_2\text{, lb} = \frac{\text{(MG)(Dosage, mg/L)(8.34 lb/gal)(100\%)}}{\text{Available chlorine, \%}}$$

$$Ca(OCl)_2\text{, lb} = \frac{\text{(0.05496 MG)(50.0 mg/L)(8.34 lb/gal)(100\%)}}{\text{65.0\%}} = 35.3 \text{ lb of } Ca(OCl)_2$$

2. **d.** Fecal contamination

   Reference: *Water System Operations (WSO) Series (AWWA)*

3. **a.** Chloroform

   Reference: *Water System Operations (WSO) Series (AWWA)*

4. **c.** 98.39% meter accuracy

   Reference: *Math for Distribution System Operators (AWWA)*

   First, convert cubic feet to gallons.

   $$\text{Number of gallons} = (118.4 \text{ ft}^3)(7.48 \text{ gal/ft}^3) = 885.632 \text{ gal}$$

   $$\text{Equation : Meter accuracy, \%} = \frac{\text{(Meter reading, gal)(100\%)}}{\text{Actual volume, gal}}$$

   $$\text{Meter accuracy, \%} = \frac{\text{(885.632 gal)(100\%)}}{\text{900.1 gal}} = 98.39\% \text{ meter accuracy}$$

5. **d.** Tastes and odors

   Reference: *Water System Operations (WSO) Series (AWWA)*

6. **c.** reduction in oxygen

   Reference: *Water Distribution Operator Training Handbook (AWWA)*

7. **a.** 2 mg/L

   Reference: *Water Distribution Operator Training Handbook (AWWA)*

8. **b.** 4-log, viruses

   Reference: *Water Distribution Operator Training Handbook (AWWA)*

9. **d.** 2.52 mg/L

   Reference: *Math for Distribution System Operators (AWWA)*

   First, convert cubic feet per second to mgd.

   $$\text{mgd} = \frac{(23.4 \text{ ft}^3\text{/s})(7.48 \text{ gal/ft}^3)(86,400 \text{ s/d})}{1,000,000/1\text{M}} = 15.123 \text{ mgd}$$

   $$\text{Equation: lb/d} = \text{(mgd)(Dosage, mg/L)(8.34 lb/d)}$$

   Rearrange to solve for dosage.

   $$\text{Dosage, mg/L} = \frac{\text{Number of lb/d of chlorine}}{\text{(mgd)(8.34 lb/gal)}}$$

   $$\text{Dosage, mg/L} = \frac{\text{318 lb/d}}{\text{(15.123 mgd)(8.34 lb/gal)}} = 2.5213 \text{ mg/L, round to 2.52mg/L}$$

10. **c.** 26 gal of NaOCl

Reference: *Math for Distribution System Operators (AWWA)*

First, calculate the number of gallons in the pipeline.

Equation: Number of gal = (0.785)(Diameter, ft)$^2$(Length, ft)(7.48 gal/ft$^3$)

Number of gal = (0.785)(2.5 ft)(2.5 ft)(4,408 ft)(7.48 gal/ft$^3$) = 161,768 gal

Convert gallons to MG.

Number of MG = (161,768 gal)(1 M/1,000,000) = 0.161768 MG

Next, determine the number of pounds of sodium hypochlorite (NaOCl) needed using the "pounds" formula.

NaOCl, lb = (MG)(Dosage, mg/L)(8.34 lb/gal)

NaOCl, lb = (0.161768 MG)(25 mg/L)(8.34 lb/gal) = 33.729 lb

Last, determine the number of gallons needed.

$$\text{NaOCl, gal} = \frac{(\text{Number of lb, NaOCl})(100\%)}{(\text{NaOCl, lb/gal})(6.0\%)}$$

$$\text{NaOCl, gal} = \frac{(33.729\ \text{lb})(100\%)}{(10.34\ \text{lb/gal})(12.5\%)} = 26.096\ \text{gal, round to 26 gal of NaOCl}$$

11. **b.** lower concentrations of THMs, ineffective, inactivating *Giardia*

Reference: *Water System Operations (WSO) Series (AWWA)*

12. **b.** 151 tablets of $Ca(OCl)_2$

Reference: *Math for Distribution System Operators (AWWA)*

First, determine the volume of the pipeline in million gallons.

$$\text{Pipeline, MG} = \frac{(0.785)(\text{Diameter, ft})^2(\text{Length, ft})(7.48\ \text{gal/ft}^3)}{1{,}000{,}000/\text{M}}$$

$$\text{Pipeline, MG} = \frac{(0.785)(3.0\ \text{ft})(3.0\ \text{ft})(1{,}865\ \text{ft})(7.48\ \text{gal/ft}^3)}{1{,}000{,}000/\text{M}} = 0.09856\ \text{MG}$$

Next, determine the number of pounds of calcium hypochlorite needed.

$$Ca(OCl)_2\text{, lb} = \frac{(\text{MG})(\text{Dosage, mg/L})(8.34\ \text{lb/gal})}{\%\ \text{purity}}$$

$$Ca(OCl)_2\text{, lb} = \frac{(0.09856\ \text{MG})(50.0\ \text{mg/L})(8.34\ \text{lb/gal})}{65.0\%/100\%} = 63.23\ \text{lb}$$

Last, find the number of tablets needed.

$$\text{Tablets, number} = \frac{63.23\ \text{lb}}{0.42\ \text{lb per tablet}} = 150.55\ \text{tablets, round to 151 tablets of } Ca(OCl)_2$$

13. **a.** 2.96 mg/L

Reference: *Water System Operations (WSO) Series (AWWA)*

First, determine the number of pounds of chlorine used.

$$\text{Equation: Chlorine, lb} = \frac{(\text{Hypochlorite, gal})(\text{Hypochlorite, \%})(8.34\ \text{lb/gal})}{100\%}$$

$$\text{Chlorine, lb} = \frac{(22.6\ \text{gal})(9.5\%)(8.34\ \text{lb/gal})}{100\%} = 17.906\ \text{lb}$$

Next, convert gallons of water to million gallons.

$$\text{MG} = (725{,}000\ \text{gal})(1\text{M}/1{,}000{,}000) = 0.725\ \text{MG}$$

Lastly, determine the actual chlorine dosage.

$$\text{Equation: Dosage, mg/L} = \frac{\text{Chlorine, lb}}{(\text{Volume, MG})(8.34\ \text{lb/gal})}$$

$$\text{Dosage, mg/L} = \frac{17.906\ \text{lb}}{(0.725\ \text{MG})(8.34\ \text{lb/gal})} = 2.96\ \text{mg/L}$$

14. **b.** 0.05–0.2 mg/L

Reference: *Water System Operations (WSO) Series (AWWA)*

15. **a.** 50.1 mg/L of $Ca(OCl)_2$

Reference: *Math for Distribution System Operators (AWWA)*

First, convert the diameter of the pipe from inches to feet.

Number of ft = (24.0 in.)(1 ft/12 in.) = 2.00 ft

Next, determine the number of gallons in the pipeline.

Number of gal = (0.785)(2.0 ft)(2.0 ft)(1,248 ft)(7.48 gal/$ft^3$) = 29,313 gal

Then convert gallons to million gallons.

Number of MG = (29,313 gal)(1 M/1,000,000) = 0.029313 MG

Last, solve for the dosage of calcium hypochlorite using the "pounds" formula.

$Ca(OCl)_2$, lb = (MG)(Dosage, mg/L)(8.34 lb/gal)

$$Ca(OCl)_2\ \text{dosage, mg/L} = \frac{Ca(OCl)_2\text{, lb}}{(\text{MG})(8.34\ \text{lb/gal})}$$

$$Ca(OCl)_2\ \text{dosage, mg/L} = \frac{12.25\ \text{lb, } Ca(OCl)_2}{(0.029313\ \text{MG})(8.34\ \text{lb/gal})}$$

$Ca(OCl)_2$ dosage, mg/L = 50.11 mg/L, round to 50.1 mg/L of $Ca(OCl)_2$

16. **d.** 64 tablets

Reference: *Math for Distribution System Operators (AWWA)*

First, determine the volume of the pipelines in million gallons.

$$\text{Pipe, MG} = \frac{(0.785)(\text{Diameter, ft})^2(\text{Length, ft})(7.48\ \text{gal/ft}^3)}{1{,}000{,}000/\text{M}}$$

$$\text{Pipe A, MG} = \frac{(0.785)(2.00\ \text{ft})(2.00\ \text{ft})(1{,}015\ \text{ft})(7.48\ \text{gal/ft}^3)}{1{,}000{,}000/\text{M}} = 0.02384\ \text{MG}$$

$$\text{Pipe B, MG} = \frac{(0.785)(1.50\ \text{ft})(1.50\ \text{ft})(1{,}347\ \text{ft})(7.48\ \text{gal/ft}^3)}{1{,}000{,}000/\text{M}} = 0.01780\ \text{MG}$$

Next, add the number of MG in the two pipes.

Total MG in both pipes = 0.02384 MG + 0.01780 MG = 0.04164 MG

Next, determine the number of pounds of calcium hypochlorite needed.

$$Ca(OCl)_2\text{, lb} = \frac{(\text{MG})(\text{Dosage, mg/L})(8.34\ \text{lb/gal})}{\text{Percentage available chlorine}}$$

$$Ca(OCl)_2\text{, lb} = \frac{(0.04164\ \text{MG})(50.0\ \text{mg/L})(8.34\ \text{lb/gal})}{65.0\%/100\%\ \text{available chlorine}} = 26.71\ \text{lb}$$

Last, find the number of tablets needed.

$$\text{Number of tablets} = \frac{26.71\ \text{lb}}{0.42\ \text{lb per tablet}} = 63.6\text{, round to 64 tablets}$$

17. **c.** Heterotrophic plate counts

Reference: *Water Distribution Operator Training Handbook (AWWA)*

18. **b.** control corrosiveness of the finished water

Reference: *Water Distribution Operator Training Handbook (AWWA)*

19. **d.** 84 gal of NaOCl

Reference: *Math for Distribution System Operators (AWWA)*

First, determine the capacity in gallons of the storage tank.

Equation: Volume, gal = $(0.785)(\text{Diameter, ft})^2(\text{Height, ft})(7.48\ \text{gal/ft}^3)(10\%\ \text{full}/100\%)$

Volume, gal = $(0.785)(105\ \text{ft})(105\ \text{ft})(32.5\ \text{ft})(7.48\ \text{gal/ft}^3)(10\%\ \text{full}/100\%)$ = 210,394 gal

Next, convert gallons to million gallons.

Number of MG = (210,394 gal)(1 M/1,000,000) = 0.210394 MG

Next, find the number of pounds of sodium hypochlorite needed using the "pounds" equation. Note: Drop the day on both sides of the equation as it is not needed in this case.

Equation: Chlorine, lb = (MG)(Dosage, mg/L)(8.34 lb/gal)

Chlorine, lb = (0.210394 MG)(50 mg/L)(8.34 lb/gal) = 87.734 lb

$$\text{NaOCl gal} = \frac{(\text{Chlorine, lb})(100\%)}{(8.34\ \text{lb/gal})(\text{Hypochlorite, }\%)}$$

$$\text{NaOCl gal} = \frac{(87.734\ \text{lb})(100\%)}{(8.34\ \text{lb/gal})(12.5\%)} = 84.16\ \text{gal, round to 84 gal of NaOCl}$$

20. **b.** Sediment deposition

Reference: *Water System Operations (WSO) Series (AWWA)*

21. **a.** Size of the water system

Reference: *Water System Operations (WSO) Series (AWWA)*

22. **c.** 0.64 mg/L, 0.48 mg/L, *Giardia lamblia*

Reference: *Water Distribution Operator Training Handbook (AWWA)*

23. **d.** All of the above

Reference: *Water System Operations (WSO) Series (AWWA)*

24. **b.** high amounts, children

Reference: *Water Distribution Operator Training Handbook (AWWA)*

25. **a.** 66 mL/min

Reference: *Math for Distribution System Operators (AWWA)*

First, determine the number of gallons that was pumped.

Equation: Tank volume, gal = (0.785)(Diameter, ft)$^2$(Level Drop, ft)(7.48 gal/ft$^3$)

Tank volume, gal = (0.785)(3.5 ft)(3.5 ft)(0.35 ft)(7.48 gal/ft$^3$) = 25.175 gal

Next, find the number of minutes in 24 hours.

Number of min = (24.0 hours)(60 min/h) = 1,440 min

Last, determine the number of milliliters per minute.

$$\text{Number of gpm} = \frac{(25.175\text{ gal})(3785\text{ mL/gal})}{1,440\text{ min}} = 66.17\text{ mL/min, round to 66 mL/min}$$

26. **d.** 28.7 lb/day of chlorine

Reference: *Math for Distribution System Operators (AWWA)*

First, convert the pumping rate to mgd.

$$\text{Equation : mgd} = \frac{(\text{pumping rate, gpm})(1,440\text{ min/d})}{1,000,000/\text{M}}$$

Substitute and solve:

$$\text{Equation : mgd} = \frac{(735\text{ gpm})(1,440\text{ min/d})}{1,000,000/\text{M}} = 1.0584\text{ mgd}$$

Next, use the "pounds" equation to solve the problem.

Equation: Chlorine, lb/d = (mgd)(Dosage, mg/L)(8.34 lb/gal)

Chlorine, lb/d = (1.0584 mgd)(3.25 mg/L)(8.34 lb/gal) = 28.7 lb/d of chlorine

27. **a.** 30 min

Reference: *Water Distribution Operator Training Handbook (AWWA)*

28. **c.** auto decomposition, ammonia

Reference: *Water Distribution Operator Training Handbook (AWWA)*

29. **d.** Water temperature approaching ambient temperature

Reference: *Water System Operations (WSO) Series (AWWA)*

30. **c.** cross-connection

Reference: *Water System Operations (WSO) Series (AWWA)*

31. **a.** coliform, total coliform

Reference: *Water Distribution Operator Training Handbook (AWWA)*

32. **d.** Both b and c

Reference: *Water Distribution Operator Training Handbook (AWWA)*

33. **d.** 417 lb of HTH

Reference: *Math for Distribution System Operators (AWWA)*

$$\text{Equation: HTH, percent solution} = \frac{(\text{lb HTH})(100\%)}{(\text{Number of gal})(8.34\ \text{lb/gal})}$$

Rearrange the equation to solve for pounds of HTH:

HTH, lb = (% solution)(Number of gal)(8.34 lb/gal)/100%

HTH, lb = (10.0% solution)(500 gal)(8.34 lb/gal)/100% =

HTH, lb = 417 lb of HTH

34. **b.** monthly

Reference: *Water Distribution Operator Training Handbook (AWWA)*

35. **b.** not understanding requirements.

Reference: *Water Distribution Operator Training Handbook (AWWA)*

36. **c.** 28 lb of sodium hypochlorite

Reference: *Math for Distribution System Operators (AWWA)*

First, convert gallons to million gallons.

Number of MG = (8,500 gal)(1 M/1,000,000) = 0.0085 MG

Then, determine the pounds of calcium hypochlorite needed.

$$\text{Equation: NaOCl, lb} = \frac{(\text{MG})(\text{Dosage, mg/L})(8.34\ \text{lb/gal})}{\%\ \text{Available chlorine}}$$

$$\text{NaOCl, lb} = \frac{(0.0085\ \text{MG})(50\ \text{mg/L})(8.34\ \text{lb/gal})}{12.5\%/100\%}$$

NaOCl, lb = 28.356 lb

37. **a.** chloramination, HAA5

Reference: *Water Distribution Operator Training Handbook (AWWA)*

38. **a.** 1–3 ft

Reference: *Water Distribution Operator Training Handbook (AWWA)*

39. **a.** 2.8% soda ash slurry

Reference: *Math for Distribution System Operators (AWWA)*

$$\text{Percentage of soda ash slurry} = \frac{(\text{Soda ash, lb})(100\%)}{\text{Soda ash, lb} + (8.34\ \text{lb/gal})(\text{Number gallons of water})}$$

Substitution:

$$\text{Percentage of soda ash slurry} = \frac{(18\ \text{lb})(100\%)}{18\ \text{lb} + (8.34\ \text{lb/gal})(75\ \text{gal})}$$

$$\frac{(18\text{ lb})(100\%)}{18\text{ lb} + 625.5\text{ lb}} = \frac{(18\text{ lb})(100\%)}{643.5\text{ lb}} = 2.8\%\text{ soda ash slurry}$$

40. **d.** Chlorine residual

Reference: *Water System Operations (WSO) Series (AWWA)*

41. **c.** Atom

42. **a.** 29.0 lb of soda ash

Reference: *Math for Distribution System Operators (AWWA)*

This type of problem in which there is one unknown can be set up as a ratio.

Let x = unknown (pounds of soda ash). The equation is:

$$\frac{x\text{ lb, Soda ash}}{\text{Number of gal}_2} = \frac{\text{Number of lb, Soda ash}}{\text{Number of gal}_1}$$

$$\frac{x\text{ lb, Soda ash}}{100\text{ gal}} = \frac{45\text{ lb, soda ash}}{155\text{ gal}}$$

$$x\text{ lb, soda ash} = \frac{(45\text{ lb, soda ash})(100\text{ gal})}{155\text{ gal}} = 29.03\text{ lb, round to 29.0 lb of soda ash}$$

43. **d.** 1, more than 5%

Reference: *Water System Operations (WSO) Series (AWWA)*

## Security, Safety, Administrative Procedures, and Public Interactions

1. **c.** nonalloyed carbon steel

Reference: *Water System Operations (WSO) Series (AWWA)*

2. **d.** The employer

Reference: *Water System Operations (WSO) Series (AWWA)*

3. **a.** soil has been disturbed such that a concrete block will not work

Reference: *Water System Operations (WSO) Series (AWWA)*

4. **b.** Severe diarrhea

Reference: *Water System Operations (WSO) Series (AWWA)*

5. **d.** Both a and b

6. **c.** refer to local utility policy

Reference: *Water Distribution Operator Training Handbook (AWWA)*

7. **b.** encouraged, have never been

Reference: *Water System Operations (WSO) Series (AWWA)*

8. **a.** taste, color, and odor

Reference: *Water Distribution Operator Training Handbook (AWWA)*

9. **a.** Physical threat

Reference: *Water System Operations (WSO) Series (AWWA)*

10. **b.** vehicles

Reference: *Water System Operations (WSO) Series (AWWA)*

11. **a.** side effects that endanger public health.
    Reference: *Water System Operations (WSO) Series (AWWA)*
12. **c.** SDWA, 1996 Amendment
    Reference: *Water System Operations (WSO) Series (AWWA)*
13. **b.** USEPA
    Reference: *Water System Operations (WSO) Series (AWWA)*
14. **a.** Terrorism
    Reference: *Water Distribution Operator Training Handbook (AWWA)*
15. **c.** PVC or polyethylene
    Reference: *Water Distribution Operator Training Handbook (AWWA)*
16. **a.** lane width
    Reference: *Water System Operations (WSO) Series (AWWA)*
17. **a.** carbon dioxide or dry powder, water or soda-acid
    Reference: *Water System Operations (WSO) Series (AWWA)*
18. **a.** drag shield
    Reference: *Water System Operations (WSO) Series (AWWA)*
19. **a.** Bioterrorism, 3,300
    Reference: *Water System Operations (WSO) Series (AWWA)*
20. **a.** governor; the particular state's EPA
    Reference: *Water Distribution Operator Training Handbook (AWWA)*
21. **a.** hard clay
    Reference: *Water System Operations (WSO) Series (AWWA)*
22. **b.** IESWTR
    Reference: *Water Distribution Operator Training Handbook (AWWA)*
23. **b.** Firm clays and tills with low moisture content
    Reference: *Water System Operations (WSO) Series (AWWA)*
24. **a.** steep
    Reference: *Water System Operations (WSO) Series (AWWA)*
25. **d.** Both b and c
    Reference: *Water Distribution Operator Training Handbook (AWWA)*
26. **c.** utilities
    Reference: *Water System Operations (WSO) Series (AWWA)*
27. **b.** develop standards
    Reference: *Water System Operations (WSO) Series (AWWA)*
28. **d.** All of the above
    Reference: *Water Distribution Operator Training Handbook (AWWA)*

29. **a.** access control

Reference: *Water Distribution Operator Training Handbook (AWWA)*

30. **b.** Card-reader systems

Reference: *Water System Operations (WSO) Series (AWWA)*

31. **a.** 12 months

Reference: *Water Distribution Operator Training Handbook (AWWA)*

This page intentionally blank.

# IV. Water Distribution, Advanced Answers

## Distribution System Components

1. **c.** 6–8 in.

   Reference: *Water System Operations (WSO) Series (AWWA)*

2. **d.** gate valve

   Reference: *Water System Operations (WSO) Series (AWWA)*

3. **d.** Detector-check meter

   Reference: *Water System Operations (WSO) Series (AWWA)*

4. **a.** 65.6 hr

   Reference: *Math for Distribution System Operators (AWWA)*

   First, determine the capacity of the tank in gallons to the overflow.

   Equation: Tank capacity, gal = (0.785)(Diameter, ft)$^2$(Overflow, ft)(7.48 gal/ft$^3$)

   Tank capacity, gal = (0.785)(45.0 ft)(45.0 ft)(19.5 ft)(7.48 gal/ft$^3$) = 231,863 gal

   Next, determine the flow into the tank

   $$\text{Equation: Flow, gpm} = \frac{2.83(\text{Diameter})^2(\text{Length, in.})}{(\text{Height, in.})^{1/2}}$$

   Substitute known values and solve.

   $$\text{Equation: Flow, gpm} = \frac{2.83(2.00\text{ in.})(2.00\text{ in.})(31.2\text{ in.})}{(36.0\text{ in.})^{1/2}} = \frac{353.184}{6.00} = 58.864\text{ gpm}$$

   Last, find the number of hours it will take to fill the tank.

   $$\text{Tank fill, hours} = \frac{231,863\text{ gal}}{(58.864\text{ gpm})(60\text{ min/h})} = 65.649\text{ hours, round to 65.6 hours}$$

5. **c.** head to discharge, efficiency to discharge, power to discharge

   Reference: *Water System Operations (WSO) Series (AWWA)*

6. **c.** quantity, quality

   Reference: *Water System Operations (WSO) Series (AWWA)*

7. **d.** All of the above

   Reference: *Water System Operations (WSO) Series (AWWA)*

8. **a.** Steel pipe
Reference: *Water System Operations (WSO) Series (AWWA)*

9. **a.** friction, electrical energy, mechanical energy
Reference: *Water System Operations (WSO) Series (AWWA)*

10. **d.** Both a and c
Reference: *Water System Operations (WSO) Series (AWWA)*

11. **c.** Standpipes
Reference: *Water System Operations (WSO) Series (AWWA)*

12. **c.** Capacitor-start motor
Reference: *Water System Operations (WSO) Series (AWWA)*

13. **a.** off-peak charge
Reference: *Water System Operations (WSO) Series (AWWA)*

14. **c.** arch-pattern base, Minneapolis-style curb stops and boxes, service line
Reference: *Water System Operations (WSO) Series (AWWA)*

15. **d.** 491 kW
Reference: *Math for Distribution System Operators (AWWA)*

Equation: kW = (Facility's, hp)(0.746 kW/hp)

kW = (658 hp)(0.746 kW/hp) = 491 kW

16. **d.** closed, vacuum pump or ejector operated with steam, air, or water
Reference: *Water System Operations (WSO) Series (AWWA)*

17. **b.** Compound-type meter
Reference: *Water System Operations (WSO) Series (AWWA)*

18. **c.** A mechanical seal
Reference: *Water System Operations (WSO) Series (AWWA)*

19. **b.** Circular, prestressed concrete tank
Reference: *Water System Operations (WSO) Series (AWWA)*

20. **b.** Pneumatic actuators
Reference: *Water System Operations (WSO) Series (AWWA)*

21. **a.** Pressure
Reference: *Water System Operations (WSO) Series (AWWA)*

22. **b.** Simplex
Reference: *Water Distribution Operator Training Handbook (AWWA)*

23. **b.** Electric actuators
Reference: *Water System Operations (WSO) Series (AWWA)*

24. **d.** Both b and c
Reference: *Water Distribution Operator Training Handbook (AWWA)*

25. **b.** Fire, 50%
Reference: *Water System Operations (WSO) Series (AWWA)*

26. **c.** Compound meter

Reference: *Water System Operations (WSO) Series (AWWA)*

27. **b.** 10 mg/L

Reference: *Water System Operations (WSO) Series (AWWA)*

28. **b.** system is being stressed.

Reference: *Water Distribution Operator Training Handbook (AWWA)*

29. **b.** friction, increase in flow rate

Reference: *Water System Operations (WSO) Series (AWWA)*

30. **a.** Reliability

Reference: *Water System Operations (WSO) Series (AWWA)*

31. **a.** 146 Hp

Reference: *Math for Distribution System Operators (AWWA)*

First, determine the pumping head for this pump.

Pumping head = Elevation of water plant − Elevation of pump

Pumping head = 605.11 ft − 297.45 ft = 307.66 ft

Next, calculate the friction loss in the pipeline.

$$\text{Friction loss} = \frac{(3,108 \text{ ft})(1.14 \text{ ft})}{1000 \text{ ft}} = 3.54 \text{ ft}$$

Now, calculate the total head

Total dynamic head = Suction lift + Pumping head + Friction loss + Velocity head

TDH = 4.52 ft + 307.66 ft + 3.54 + 2.48 ft

TDH = 318.2 ft

Last, determine the required horsepower of the pump.

$$\text{Hp} = \frac{(\text{gpm})(\text{TDH})}{(3,960)(\text{Pump efficiency})(\text{Motor efficiency})}$$

$$\text{Hp} = \frac{(1,250 \text{ gpm})(318.2 \text{ ft})}{(3,960)(78.6\%/100\%)(87.6\%/100\%)}$$

Hp = 145.87 Hp, round up to 146 Hp

32. **a.** Hydraulic actuators

Reference: *Water System Operations (WSO) Series (AWWA)*

33. **d.** Both b and c

Reference: *Water System Operations (WSO) Series (AWWA)*

34. **b.** Bypass valve

Reference: *Water System Operations (WSO) Series (AWWA)*

35. **c.** Vertical turbine pump

Reference: *Water System Operations (WSO) Series (AWWA)*

36. **b.** tastes and odors, NSF International standards

Reference: *Water System Operations (WSO) Series (AWWA)*

37. **a.** Valve closes horizontally.

Reference: *Water System Operations (WSO) Series (AWWA)*

38. **b.** 2,290,000 gal

Reference: *Math for Distribution System Operators (AWWA)*

First, determine the water level in feet that is in the storage tank.

Equation:

$$\text{Current process reading} = \frac{(\text{Live signal, mA} - 4\text{ mA offset})(\text{Maximum capacity})}{16\text{ mA span}}$$

Substitute known values and solve:

$$\text{Storage tank level, ft} = \frac{(14.88\text{ mA} - 4\text{mA offset})(29.2\text{ ft maximum level})}{16\text{ mA span}}$$

$$\text{Storage tank level, ft} = \frac{(10.88\text{ mA})(29.2\text{ ft maximum level})}{16\text{ mA span}}$$

Storage tank level, ft = 19.856 ft

Now, determine the number of gallons in the storage tank.

Storage tank, gal = $\pi$(Radius, ft)$^2$(Level, ft)(7.48 gal/ft$^3$)

Storage tank, gal = $\pi$(70.0 ft)$^2$(19.856 ft)(7.48 gal/ft$^3$)

Storage tank, gal = 3.14(70.0 ft)(70.0 ft)(19.856 ft)(7.48 gal/ft$^3$)

Storage tank, gal = 2,285,173 gal, round to 2,290,000 gal

39. **b.** proportional to the square of the velocity

Reference: *Water Distribution Operator Training Handbook (AWWA)*

40. **d.** Pulse-duration modulation

Reference: *Water System Operations (WSO) Series (AWWA)*

41. **c.** Grooved joints

Reference: *Water System Operations (WSO) Series (AWWA)*

42. **c.** 8–12 in.

Reference: *Water System Operations (WSO) Series (AWWA)*

43. **c.** Horizontal gate valve

Reference: *Water System Operations (WSO) Series (AWWA)*

44. **d.** Radial-flow pump

45. **a.** Locked rotor current

Reference: *Water System Operations (WSO) Series (AWWA)*

46. **a.** Elevated storage tanks
Reference: *Water System Operations (WSO) Series (AWWA)*

47. **b.** Water losses and energy usage
Reference: *Water System Operations (WSO) Series (AWWA)*

48. **c.** 2–8 in.
Reference: *Water System Operations (WSO) Series (AWWA)*

49. **b.** Motor efficiency times pump efficiency
Reference: *Water System Operations (WSO) Series (AWWA)*

50. **d.** Air-relief valve
Reference: *Water System Operations (WSO) Series (AWWA)*

51. **d.** Float mechanism
Reference: *Water System Operations (WSO) Series (AWWA)*

52. **c.** Pulse-duration modulation and variable frequency
Reference: *Water System Operations (WSO) Series (AWWA)*

53. **a.** only of the closed design
Reference: *Water System Operations (WSO) Series (AWWA)*

54. **a.** PVC pipe
Reference: *Water System Operations (WSO) Series (AWWA)*

55. **c.** The valve closes with the water pressure against the seat.
Reference: *Water System Operations (WSO) Series (AWWA)*

56. **b.** stator, current-carrying coils
Reference: *Water System Operations (WSO) Series (AWWA)*

57. **c.** in cold climates
Reference: *Water System Operations (WSO) Series (AWWA)*

58. **d.** Globe valve
Reference: *Water System Operations (WSO) Series (AWWA)*

59. **b.** low velocity, high pressure, shape of the volute or shape of the diffuser vanes
Reference: *Water System Operations (WSO) Series (AWWA)*

60. **a.** Velocity, vertical
Reference: *Water System Operations (WSO) Series (AWWA)*

61. **c.** polybutylene pipe
Reference: *Water System Operations (WSO) Series (AWWA)*

62. **b.** Steel pipe
Reference: *Water System Operations (WSO) Series (AWWA)*

63. **b.** 106 mhp

Reference: *Math for Distribution System Operators (AWWA)*

First, convert million gallons per day to gallons per minute.

gpm = (4.22 mgd)(1,000,000/1 M)(1 day/1,440 min) = 2,931 gpm

Next, convert the percentage motor and pump efficiencies to decimal form by dividing by 100%

Motor efficiency = 88.6%/100% = 0.886

Pump efficiency = 79.3%/100% = 0.793

The equation for determining the motor horsepower with the given data is different than the problem above.

$$\text{Equation : mhp} = \frac{(\text{Flow, gpm})(\text{TH, ft})}{(3,960)(\text{Motor efficiency})(\text{Pump efficiency})}$$

Note: The number 3,960 is a constant.

$$\text{mhp} = \frac{(2,931\ \text{gpm})(101\ \text{ft})}{(3,960)(0.886\ \text{Motor efficiency})(0.793\ \text{Pump efficiency})} = 106.4\ \text{mhp, round to 106 mhp}$$

64. **c.** Hazen–Williams formula

Reference: *Water System Operations (WSO) Series (AWWA)*

65. **c.** Feedback control

Reference: *Water System Operations (WSO) Series (AWWA)*

66. **a.** hydraulically, 20,000 hours

Reference: *Water System Operations (WSO) Series (AWWA)*

67. **a.** Fiberglass pipe

Reference: *Water System Operations (WSO) Series (AWWA)*

68. **b.** Current and voltage

Reference: *Water System Operations (WSO) Series (AWWA)*

69. **a.** gate valve

Reference: *Water System Operations (WSO) Series (AWWA)*

70. **a.** nothing exposed inside that can be damaged by the ice

Reference: *Water System Operations (WSO) Series (AWWA)*

71. **a.** rotor and stator

Reference: *Water System Operations (WSO) Series (AWWA)*

72. **a.** 3–15 psig, 4-20 mA direct current (DC)

Reference: *Water Distribution Operator Training Handbook (AWWA)*

73. **b.** a key, key nut

Reference: *Water System Operations (WSO) Series (AWWA)*

74. **a.** state coordinate system

Reference: *Water System Operations (WSO) Series (AWWA)*

75. **b.** Feed forward control

Reference: *Water System Operations (WSO) Series (AWWA)*

76. **a.** relatively accurate, difficult to maintain, little friction loss
Reference: *Water System Operations (WSO) Series (AWWA)*

77. **c.** Magnetic starters
Reference: *Water System Operations (WSO) Series (AWWA)*

78. **a.** Transducer
Reference: *Water System Operations (WSO) Series (AWWA)*

79. **b.** main lines
Reference: *Water System Operations (WSO) Series (AWWA)*

80. **a.** Needle valve
Reference: *Water System Operations (WSO) Series (AWWA)*

81. **a.** Feedback control
Reference: *Water System Operations (WSO) Series (AWWA)*

82. **d.** sewage
Reference: *Water System Operations (WSO) Series (AWWA)*

83. **b.** plan and profile drawings, engineering drawings, section lines or property lines
Reference: *Water System Operations (WSO) Series (AWWA)*

84. **a.** Close-coupled
Reference: *Water System Operations (WSO) Series (AWWA)*

85. **a.** velocity pump
Reference: *Water System Operations (WSO) Series (AWWA)*

86. **a.** 2–5, narrower
Reference: *Water System Operations (WSO) Series (AWWA)*

87. **c.** steel or concrete pipe
Reference: *Water System Operations (WSO) Series (AWWA)*

88. **d.** 6 ft
Reference: *Water System Operations (WSO) Series (AWWA)*

89. **c.** Magnesium
Reference: *Water System Operations (WSO) Series (AWWA)*

90. **a.** 0.010–0.020 in.
Reference: *Water System Operations (WSO) Series (AWWA)*

91. **d.** High repair costs
Reference: *Water System Operations (WSO) Series (AWWA)*

## Equipment Installation, Operation, and Maintenance

1. **d.** ball, plug
Reference: *Water System Operations (WSO) Series (AWWA)*

2. **a.** Sulfur dioxide
Reference: *Water System Operations (WSO) Series (AWWA)*

3. **c.** Zinc or magnesium

   Reference: *Water System Operations (WSO) Series (AWWA)*

4. **b.** As-built plans

   Reference: *Water System Operations (WSO) Series (AWWA)*

5. **b.** Wet-top hydrants

   Reference: *Water System Operations (WSO) Series (AWWA)*

6. **c.** underground storage facilities

   Reference: *Water System Operations (WSO) Series (AWWA)*

7. **a.** polypigs

   Reference: *Water System Operations (WSO) Series (AWWA)*

8. **d.** two points, voltage, transducer

   Reference: *Water System Operations (WSO) Series (AWWA)*

9. **b.** 6 in.

   Reference: *Water Distribution Operator Training Handbook (AWWA)*

10. **b.** 92.7 lb of lime

    Reference: *Math for Distribution System Operators (AWWA)*

Let x = the number of pounds of lime needed.

$$10.0\%\text{ lime} = \frac{(\text{x lb})(100\%)}{(\text{x lb}) + (8.34\text{ lb/gal})(100\text{ gal})}$$

To solve this problem, we have to use a little more complex algebra than what we have been using so far. It is not that difficult, just follow the steps, and always remember that whatever we do to one side of the equation, we have to do to the other side.

First, multiply both sides by (x lb) + (8.34 lb/gal)(100 gal)

$$(\text{x lb})(10.0\%\text{ lime}) + (8.34\text{ lb/gal})(100\text{ gal})(10.0\%\text{ lime}) = \frac{(\text{x lb})(100\%)\{(\text{x lb}) + (8.34\text{ lb/gal})(100\text{ gal})\}}{\{(\text{x lb}) + (8.34\text{ lb/gal})(100\text{ gal})\}}$$

The terms in the { } will cancel each other in the numerator (top) and the denominator (bottom) to get the following:

$$(\text{x lb})(10.0\%\text{ lime}) + (8.34\text{ lb/gal})(100\text{ gal})(10.0\%\text{ lime}) = (\text{x lb})(100\%)$$

Next, divide both sides by 100%

$$\frac{(\text{x lb})(10.0\%\text{ lime})}{100\%} + \frac{(8.34\text{ lb/gal})(100\text{ gal})(10.0\%\text{ lime})}{100\%} = \frac{(\text{x lb})(100\%)}{100\%}$$

The 100% will cancel on the right side of the equation to get the following:

$$\frac{(\text{x lb})(10.0\%\text{ lime})}{100\%} + \frac{(8.34\text{ lb/gal})(100\text{ gal})(10.0\%\text{ lime})}{100\%} = \text{x lb}$$

Next, reduce the problem by solving the multiplication and division of

(8.34 lb/gal)(100 gal)(10.0% lime)/100%

and division of (x lb)(10.0% lime)/100%

0.1x lb of lime + 83.4 lb of lime = x lb

Next, subtract 0.1x lb of lime from both sides to get the following:

0.1x lb of lime − 0.1x lb of lime + 83.4 lb of lime = x lb − 0.1 lb of lime

Add the like terms (the x terms) to get the following:

83.4 lb = 0.9x lb of lime

Last, divide both sides by 0.9.

$$\frac{83.4 \text{ lb}}{0.9} = \frac{0.9 \times \text{lb of lime}}{0.9}$$

x lb of lime = 92.67 lb of lime, round to 92.7 lb of lime

11. **c.** At least 16 hours
    Reference: *Water Distribution Operator Training Handbook (AWWA)*
12. **b.** closed-cell Styrofoam insulation board
    Reference: *Water System Operations (WSO) Series (AWWA)*
13. **c.** woven polyester
    Reference: *Water Distribution Operator Training Handbook (AWWA)*
14. **c.** unnecessary damage
    Reference: *Water System Operations (WSO) Series (AWWA)*
15. **b.** compound, molecule
    Reference: *Water System Operations (WSO) Series (AWWA)*
16. **c.** pump control, a vacuum breaker
    Reference: *Water System Operations (WSO) Series (AWWA)*
17. **b.** Impressed current cathodic, direct, cathodic
    Reference: *Water System Operations (WSO) Series (AWWA)*
18. **b.** Deep well pump
    Reference: *Water System Operations (WSO) Series (AWWA)*
19. **d.** Calcium
    Reference: *Water System Operations (WSO) Series (AWWA)*
20. **c.** Sodium sulfite
    Reference: *Water System Operations (WSO) Series (AWWA)*
21. **c.** Pulse-duration modulation
    Reference: *Water System Operations (WSO) Series (AWWA)*
22. **b.** Clays, tills, and silts
    Reference: *Water System Operations (WSO) Series (AWWA)*
23. **b.** The excavation
    Reference: *Water System Operations (WSO) Series (AWWA)*
24. **c.** pipe joint or a damaged pipe, valve or service connection
    Reference: *Water System Operations (WSO) Series (AWWA)*

25. **c.** Tablet method
    Reference: *Water System Operations (WSO) Series (AWWA)*

26. **d.** 24 hours
    Reference: *Water System Operations (WSO) Series (AWWA)*

27. **b.** 13.9 ft
    Reference: *Math for Distribution System Operators (AWWA)*

    Write the equation, arranging it to solve for the unknown, drawdown.

    Equation: Drawdown, ft = Well yield, gpm/Specific yield, gpm/ft

    Drawdown, ft = 206 gpm/14.8 gmp/ft = 13.919 ft, round to 13.9 ft

28. **b.** Dry-top barrel hydrants
    Reference: *Water System Operations (WSO) Series (AWWA)*

29. **c.** Inserting valve
    Reference: *Water System Operations (WSO) Series (AWWA)*

30. **b.** well-rounded, four to five, larger
    Reference: *Water System Operations (WSO) Series (AWWA)*

31. **c.** 2 ft
    Reference: *Water System Operations (WSO) Series (AWWA)*

32. **b.** inversely proportional
    Reference: *Water System Operations (WSO) Series (AWWA)*

33. **b.** Two to three pipe volumes
    Reference: *Water Distribution Operator Training Handbook (AWWA)*

34. **a.** 38°F (3.3°C).
    Reference: *Water Distribution Operator Training Handbook (AWWA)*

35. **a.** coupling, shaft coupling, packing
    Reference: *Water System Operations (WSO) Series (AWWA)*

36. **c.** Keep pipe ends covered to prevent contamination
    Reference: *Water System Operations (WSO) Series (AWWA)*

37. **a.** External corrosion control
    Reference: *Water System Operations (WSO) Series (AWWA)*

## Disinfection Monitoring, Evaluation, Adjustment, and Laboratory Analysis/ Interpretation

1. **a.** Because microorganisms biodegrade them
   Reference: *Water System Operations (WSO) Series (AWWA)*

2. **b.** 10
   Reference: *Water Distribution Operator Training Handbook (AWWA)*

3. **b.** secondary, action levels

Reference: *Water Distribution Operator Training Handbook (AWWA)*

4. **c.** Source water treatment

Reference: *Water System Operations (WSO) Series (AWWA)*

5. **b.** 1.41 gal of sodium hypochlorite

Reference: *Math for Distribution System Operators (AWWA)*

First, convert 14 in. to feet: 14 in./12 in. per foot = 1.167 ft

Next, find the volume of the pipe using: Volume, gal = (0.785)(Diameter, ft)$^2$(Length, ft)(7.48 gal/ft$^3$)

Volume = (0.785)(1.167 ft)(1.167 ft)(1,002 ft)(7.48 gal/ft$^3$) = 8,012.7 gal

Next, find the number of MG.

MG = (16,448 gal)(1 M/1,000,000) = 0.0080127 MG

Then use the "pound" equation:

$$\text{Sodium hypochlorite, lb} = \frac{(\text{MG})(\text{Dosage, mg/L})(8.34\ \text{lb/gal})}{\%\ \text{available chlorine}/100\%}$$

$$\text{Sodium hypochlorite, lb} = \frac{(0.0080127\ \text{MG})(25.0\ \text{mg/L})(8.34\ \text{lb/gal})}{11.7\%\ \text{available chlorine}/100\%} = 14.279\ \text{lb}$$

Last, calculate the number of gallons of sodium hypochlorite.

Sodium hypochlorite, gal = 14.279 lb/10.12 lb/gal = 1.41 gal of sodium hypochlorite

6. **b.** molecular weight

Reference: *Water System Operations (WSO) Series (AWWA)*

7. **b.** 25 mg/L

Reference: *Water Distribution Operator Training Handbook (AWWA)*

8. **d.** radon

Reference: *Water System Operations (WSO) Series (AWWA)*

9. **c.** Mix 242 gal of the 12.5% solution with 583 gal of the 3.30% solution

Reference: *Water System Operations (WSO) Series (AWWA)*

Solve the problem using the dilution triangle

12.5% 2.70 2.70 parts of the 12.5% solution are required for every 9.20 parts.

6.00%

3.30% $\frac{6.50}{9.20\ \text{total parts}}$ 6.50 parts of the 3.30% solution are required for every 9.20 parts.

$$\left[\frac{2.70\ \text{parts}(825\ \text{gal})}{9.20\ \text{parts}} = 242\ \text{gallons of the 12.5\% solution}\right]$$

$$\frac{6.50\ \text{parts}\ (825\ \text{gal})}{9.20\ \text{parts}} = \frac{583\ \text{gallons of the 3.80\% solution}}{825\ \text{gallons}}$$

To make the 825 gallons of the 6.00% sodium hypochlorite solution, mix 242 gallons of the 12.5% solution with 583 gallons of the 3.30% solution.

10. **a.** TDS

Reference: *Water System Operations (WSO) Series (AWWA)*

11. **a.** 9.0 %

Reference: *Math for Distribution System Operators (AWWA)*

First, convert 3,205 gpm filter flow to mgd.

$$\frac{(3,205\ \text{gpm})(1,440\ \text{min})(1\ \text{MG})}{\text{day}\quad 1,000,000\ \text{gal}} = 4.6152\ \text{mgd}$$

Then, convert 435 gpd of the hypochlorite solution rate of flow to MGD.

$$\frac{(215\ \text{gpd})(1\ \text{MG})}{1,000,000\ \text{gal}} = 0.000215\ \text{mgd}$$

Then, using the equal dosage equations:

$$(0.000215\ \text{MGD})(x\ \text{mg/L})(10.3\ \text{lb/gal}) \quad = \quad (4.6152\ \text{MGD})(5.2\ \text{mg/L})(8.34\ \text{lb/gal})$$

$$\text{Hypochlorite, mg/L} = \frac{(4.6152\ \text{mgd})(5.2\ \text{mg/L})(8.34\ \text{lb/gal})}{(0.000215\ \text{mgd})(10.3\ \text{lb/gal})} = 90,382\ \text{mg/L}$$

Last, convert mg/L into percentage:

$$\frac{(90,382\ \text{mg/L})(1\%)}{10,000\ \text{mg/L}} = 9.0\%$$

12. **b.** 2.7 gal of NaOCl

Reference: *Water System Operations (WSO) Series (AWWA)*

First, calculate the number of gallons in the pipeline.

Equation: Number of gal = (0.785)(Diameter, ft)$^2$(Length, ft)(7.48 gal/ft$^3$)

Number of gal = (0.785)(1.50 ft)(1.50 ft)(1,251 ft)(7.48 gal/ft$^3$) = 16,528 gal

Convert gallons to MG.

Number of MG = (16,528 gal)(1 M/1,000,000) = 0.016528 MG

Next, determine the number of pounds of sodium hypochlorite (NaOCl) needed using the "pounds" formula.

NaOCl, lb = (MG)(Dosage, mg/L)(8.34 lb/gal)

NaOCl, lb = (0.016528 MG)(25 mg/L)(8.34 lb/gal) = 3.446 lb

Lastly, determine the number of gallons needed.

$$\text{NaOCl, gal} = \frac{(\text{Number of lb, NaOCl})\ (100\%)}{(\text{NaOCl, lb/gal})\ (6.0\%)}$$

$$\text{NaOCl, gal} = \frac{(3.446\ \text{lb})\ (100\%)}{(10.32\ \text{lb/gal})\ (12.5\%)} = 2.67\ \text{gal, round to 2.7 gal of NaOCl}$$

13. **c.** *Cryptosporidium*

Reference: *Water System Operations (WSO) Series (AWWA)*

14. **a.** 1.07 gal of sodium hypochlorite

Reference: *Math for Distribution System Operators (AWWA)*

First, find the volume of the pipe using:

Equation: Volume, gal = (0.785)(Diameter, ft)$^2$(Length, ft)(7.48 gal/ft$^3$)

Volume, gal = (0.785)(1.5)(1.5)(412 ft)(7.48 gal/ft$^3$) = 5,443 gal

Next, find the number of MG.

MG = (5,443 gal)(1 M/1,000,000) = 0.005443 MG

Then use the "pounds" equation and include the percentage of available chlorine:

$$\text{Sodium hypochlorite, lb} = \frac{\text{(MG)(Dosage, mg/L)(8.34 lb/gal)}}{\text{\% available chlorine/100\%}}$$

$$\text{Sodium hypochlorite, lb} = \frac{\text{(0.005443 MG)(25.0 mg/L)(8.34 lb/gal)}}{\text{11.5\% available chlorine/100\%}} = 9.868 \text{ lb}$$

Last, calculate the number of gallons of sodium hypochlorite.

Sodium hypochlorite, gal = 9.868 lb/9.25 lb/gal = 1.07 gal

15. **c.** 1.21 g/cm$^3$

Reference: *Math for Distribution System Operators (AWWA)*

First, convert the number of gallons to liters.

Polymer, L = (5.00 gal)(3.785 L/gal) = 18.925 L

Polymer, cm$^3$ = (1,000.027 cm$^3$/1 L)(18.925 L) = 18,925.5 cm$^3$

Next, convert the number of pounds to grams.

Know from conversion tables that 1 lb = 454 g and 1 L = 1,000.027 cm$^3$.

Polymer, grams = (Polymer, lb)(454 g/lb)

Polymer, grams = (50.5 lb)(454 g/lb) = 22,927 g

Density = Mass/Volume

Density = 22,927 g/18,925.5 cm$^3$ = 1.21 g/cm$^3$

16. **c.** The valence electrons

Reference: *Water System Operations (WSO) Series (AWWA)*

17. **c.** 1.82% soda ash slurry

Reference: *Water System Operations (WSO) Series (AWWA)*

First, convert ounces to pounds.

Number of lb = (12 ounces) (1 lb/16 ounces) = 0.75 lb

Total number of lb = 7 lb + 0.75 lb = 7.75 lb

$$\text{Soda ash, \%} = \frac{\text{(7.75 lb) (100\%)}}{\text{7.75 lb+ (50.0 gal) (8.34 lb/gal)}} = \frac{\text{(7.75 lb) (100\%)}}{\text{7.75 lb + 417 lb}} = \frac{\text{(7.75 lb) (100\%)}}{\text{424.75 lb}} = 1.82\%$$

18. **a.** 145 lb/day of chlorine

Reference: *Math for Distribution System Operators (AWWA)*

Equation: lb/d = (mgd)(Dosage, mg/L)(8.34 lb/d)

Chlorine, lb/d = (5.36 mgd)(3.25 mg/L)(8.34 lb/gal) = 145.28 lb/d, round to 145 lb/d of chlorine

19. **d.** breakdown of naturally occurring organic chemicals

Reference: *Water System Operations (WSO) Series (AWWA)*

20. **a.** 15 min

Reference: *Water Distribution Operator Training Handbook (AWWA)*

21. **a.** 25 mg/L, 24 hours

Reference: *Water Distribution Operator Training Handbook (AWWA)*

22. **d.** 41°F (5°C); 48 hours

Reference: *Water Distribution Operator Training Handbook (AWWA)*

23. **b.** 1.27 $g/cm^3$

Reference: *Math for Distribution System Operators (AWWA)*

Know from conversion tables that 1 lb = 454 g, and 1 L = 1,000.027 $cm^3$.

First, convert the number of liters to cubic centimeters.

Number of $cm^3$ = (1,000.027 $cm^3$/1 L)(1.559 L) = 1,559 $cm^3$

Next, convert the number of pounds to grams.

Number of grams = (Number of lb)(454 g/lb)

Number of grams = (4.35 lb)(454 g/lb) = 1,974.9 g

Density = Mass/Volume

Density = 1,974.9 g/1,559 $cm^3$ = 1.27 $g/cm^3$

24. **a.** 1% to 5%

Reference: *Water Distribution Operator Training Handbook (AWWA)*

25. **a.** 66 lb $Ca(OCl)_2$

Reference: *Math for Distribution System Operators (AWWA)*

First, find the initial amount of water to be disinfected. 10% capacity = (2.0 MG)(10%/100%) = 0.20 MG

Next, determine the number of pounds of chlorine needed by using the "pounds" formula.

$$\text{Equation: } Ca(OCl)_2\text{, lb} = \frac{(\text{MG})(\text{Dosage, mg/L})(8.34\text{ lb/gal})}{\%\text{ available chlorine}/100\%}$$

$$Ca(OCl)_2\text{, lb} = \frac{(0.20\text{ MG})(25.0\text{ mg/L})(8.34\text{ lb/gal})}{63.0\%/100\%} = 66.19\text{ lb, round to 70 lb } Ca(OCl)_2$$

26. **b.** sievert, sievert, 100

Reference: *Water System Operations (WSO) Series (AWWA)*

27. **c.** 0.101-gal $Ca(OCl)_2$

Reference: *Water System Operations (WSO) Series (AWWA)*

First, find the length (in feet) of water in the casing.

Length of water-filled casing = Depth of well – Depth of water to top of casing

Length of water-filled casing = 263 ft – 92.5 ft = 170.5 ft

Then, convert the diameter from inches to feet.

$$\text{Diameter, ft} = \frac{10.0 \text{ inches}}{12 \text{ in/ft}} = 0.833 \text{ ft}$$

Next, determine the volume in gallons of water in the well casing using the following equation:

$$\text{Volume, gal} = (0.785)\,(\text{Diameter, ft})^2\,(\text{Length, ft})\left(7.48 \text{ gal/ft}^3\right)$$

$$\text{Volume, gal} = (0.785)\,(0.833 \text{ ft})\,(0.833 \text{ ft})\,(170.5 \text{ ft})\left(7.48 \text{ gal/ft}^3\right) = 695 \text{ gal}$$

Next, determine the number of MG.

$$\text{MG} = (695 \text{ gal})\,(1\text{M}/1,000,000) = 0.000695 \text{ MG}$$

Using the "pounds" equation, calculate the number of lb of chlorine needed.

Chlorine, lb = (0.000695 MG)(50.0 mg/L)(8.34 lb/gal) = 0.2898 lb

Lastly, calculate the gallons of calcium hypochlorite solution required.

$$Ca(OCl)_2\text{, gal} = \frac{(\text{Chlorine, lb})\,(100\%)}{(10.51 \text{ lb/gal})\,(\text{Hypochlorite, \%})}$$

$$Ca(OCl)_2\text{, gal} = \frac{(0.2898 \text{ lb})\,(100\%)}{(10.42 \text{ lb/gal})\,(27.5\%)} = 0.1011 \text{ gal, round to } 0.101 \text{ gal } Ca(OCl)_2$$

28. **b.** before and after the repair site, two consecutive

Reference: *Water Distribution Operator Training Handbook (AWWA)*

29. **a.** 6 hours

Reference: *Water Distribution Operator Training Handbook (AWWA)*

30. **c.** neutralization

Reference: *Water System Operations (WSO) Series (AWWA)*

31. **d.** 8.75% strength of new solution

Reference: *Math for Distribution System Operators (AWWA)*

First, find the weight in pounds per gallon for the 11.9% solution:

Weight of 11.9% solution = (Specific gravity)(8.34 lb/gal)

Weight of 11.9% solution = (1.22 S.G.)(8.34 lb/gal) = 10.175 lb/gal

Equation: % Mixture Strength =

$$[\frac{\{\text{Soln.1, gal(lb/gal)(Avail.\%/100\%)} + \text{Soln. 2, gal(lb/gal)(Avail.\%/100\%)}\}\text{(100\% strength)}}{\text{Solution 1, gal(lbs/gal)} + \text{Solution 2, gal/(lbs/gal)}}]$$

% mixture strength =

$$\frac{245\ \text{gal}(10.175\ \text{lb/gal})(11.9\%/100\%) + 136\ \text{gal}(9.86\ \text{lb/gal})(2.90\%/100\%)(100\%)}{245\ \text{gal}(10.175\ \text{lb/gal}) + 136\ \text{gal}(9.86\ \text{lb/gal})}$$

$$\%\ \text{mixture strength} = \frac{(296.65\ \text{lb} + 38.89\ \text{lb})(100\%)}{2492.875\ \text{lb} + 1340.96\ \text{lb}}$$

$$\%\ \text{mixture strength} = \frac{(335.54\ \text{lb})(100\%)}{3,833.835\ \text{lb}} = (0.08752)(100\%) = 8.752\%$$

round to 8.75% strength of new solution

32. **c.** 1.168

Reference: *Math for Distribution System Operators (AWWA)*

First, convert kg/ft$^3$ to g/ft$^3$. 1 kg = 1,000 g

S.G. = (33.08 kg/ft$^3$)(1,000 g/kg) = 33,080 g/ft$^3$

Next, convert cubic feet by dividing by the conversion factor of 28,317 cm$^3$/ft$^3$.

$$\text{S.G.} = \frac{33,080\ \text{g/ft}^3}{28,317\ \text{cm}^3/\text{ft}^3} = 1.168$$

33. **c.** Provides more uniform disinfection at a lower level

Reference: *Water System Operations (WSO) Series (AWWA)*

34. **a.** 4.09 mg/L of soda ash

Reference: *Math for Distribution System Operators (AWWA)*

To solve this problem, we need to determine how many gallons of water was treated with the soda ash and the number of pounds of soda ash used for this treatment. First, find the number of gallons treated.

Begin by finding the number of gallons pumped to the storage tank.

Number of gal pumped = (2,905 gpm)(1,440 min) = 4,183,200 gal

Next, determine the gallons from the 4.18 ft increase in the tank's water level.

Equation: Number of gallons = (0.785)(Diameter, ft)$^2$(Level difference, ft)(7.48 gal/ft$^2$)

Number of gallons = (0.785)(120 ft)(120 ft)(4.18 ft)(7.48 gal/ft$^3$) = 353,435 gal

Because the level in the tank increased, more water was pumped into the tank than left the tank. Therefore, to find the number of gallons treated by the soda ash in 24 hours or one day, subtract the number of gallons pumped by the increase in gallons in the tank.

Total water treated, gal = 4,183,200 gal − 353,435 gal = 3,829,765 gal

Next, convert gallons to million gallons.

$$MG = \frac{3,829,765 \text{ gal}}{1,000,000/1 \text{ M}} = 3.83 \text{ MG}$$

Now, determine the soda ash usage in pounds.

Soda ash, lb = (41.2 g/min)(1,440 min/d)(1 lb/454 g) = 130.68 lb of soda ash

Last, determine the dosage using the "pounds" equation.

Equation: Soda ash, lb = (mgd)(Dosage, mg/L)(8.34 lb/gal)

Rearrange and solve for dosage in milligrams per liter.

$$\text{Dosage, mg/L} = \frac{130.68 \text{ lb/d}}{(3.83 \text{ mgd})(8.34 \text{ lb/gal})} = 4.09 \text{ mg/L of soda ash}$$

35. **d.** HAA5

    Reference: *Water System Operations (WSO) Series (AWWA)*

36. **a.** 50 mg/L, 100 mg/L

    Reference: *Water Distribution Operator Training Handbook (AWWA)*

## Security, Safety, Administrative Procedures, and Public Interactions

1. **c.** Tabun

   Reference: *Water System Operations (WSO) Series (AWWA)*

2. **a.** Teamwork

   Reference: *Water System Operations (WSO) Series (AWWA)*

3. **c.** ANSI/NSF Standard 61

   Reference: *Water Distribution Operator Training Handbook (AWWA)*

4. **c.** Trench braces, adjustable trench jacks

   Reference: *Water System Operations (WSO) Series (AWWA)*

5. **b.** The supervisor of that employee

   Reference: *Water System Operations (WSO) Series (AWWA)*

6. **a.** sodium fluoroacetate

   Reference: *Water System Operations (WSO) Series (AWWA)*

7. **c.** 600 V

8. **a.** lightweight, steel, aluminum

   Reference: *Water Distribution Operator Training Handbook (AWWA)*

9. **d.** planning.

   Reference: *Water System Operations (WSO) Series (AWWA)*

10. **d.** ANSI/AWWA Standard C651

    Reference: *Water Distribution Operator Training Handbook (AWWA)*

This page intentionally blank.

# **Appendix A:** Formulas and Conversions for Water Treatment and Distribution Exams

**Alkalinity, mg/L as $CaCO_3$** = $\frac{\text{(titrant volume, mL)(acid normality) (50,000)}}{\text{sample volume, mL}}$

**Amps** = $\frac{\text{volts}}{\text{ohms}}$

**Area of circle** = $(0.785)(\text{diameter}^2)$

**Area of circle** = $(3.14)(\text{radius}^2)$

**Area of cone (lateral area)** = $(3.14)(\text{radius})\sqrt{\text{radius}^2 + \text{height}^2}$

**Area of cone (total surface area)** = $(3.14)(\text{radius})(\text{radius} + \sqrt{\text{radius}^2 + \text{height}^2})$

**Area of cylinder**
**(total exterior surface area)** = [End #1 SA] + [End #2 SA] + [(3.14)(diameter)(height or depth)]
***Where SA = surface area***

**Area of rectangle** = (Length)(Width)

**Area of right triangle** = $\frac{\text{(Base)(Height)}}{2}$

**Average (arithmetic mean)** = $\frac{\text{sum of all terms}}{\text{number of terms}}$

**Average (geometric mean)** = $[(X_1)(X_2)(X_3)(X_4)(X_n)]^{1/n}$ *The nth root of the product of n numbers*

**Blending** = $(V_1)(C_1) + (V_2)(C_2) = (V_3)(C_3)$ *Where V = volume or flow, C = concentration or percent solution*

**Chemical feed pump setting, % stroke** = $\frac{\text{desired flow}}{\text{maximum flow}} \times 100\%$

**Chemical feed pump setting, mL/min** = $\frac{\text{(flow, mgd)(dose, mg/L)(3.785 L/gal)(1,000,000 gal/MG)}}{\text{(feed chemical density, mg/mL)(1,440 min/day)}}$

**Chemical feed pump setting, mL/min** =

$$\frac{(\text{flow, m}^3\text{/day})(\text{dose, mg/L})}{(\text{feed chemical density, g/cm}^3)(\text{active chemical, \% expressed as a decimal})(1{,}440 \text{ min/day})}$$

---

Modified with permission from Association of Boards of Certification

**Circumference of circle** $= (3.14)(\text{diameter})$

**Composite sample single portion** $= \dfrac{(\text{Instantaneous Flow})(\text{Total Sample Volume})}{(\text{Number of Portions})(\text{Average Flow})}$

**CT calculation** $= (\text{disinfectant residual concentration, mg/L})(\text{time, min})$

**Degrees Celsius** $= \dfrac{(°F - 32)}{1.8}$

**Degrees Fahrenheit** $= (°C)(1.8) + 32$

**Detention time** $= \dfrac{\text{volume}}{\text{flow}}$ ***Units must be compatible***

**Electromotive force, volts** $= (\text{current, amps})(\text{resistance, ohms})$

**Feed rate, lb/day** $= \dfrac{(\text{dosage, mg/L})(\text{flow, mgd})(8.34\ \text{lb/gal})}{\text{purity, \% expressed as a decimal}}$

**Feed rate, kg/day** $= \dfrac{(\text{dosage, mg/L})(\text{flow rate, m}^3\text{/day})}{(\text{purity, \% expressed as a decimal})(1{,}000)}$

**Feed rate (fluoride), lb/day** =

$$\frac{(\text{dosage, mg/L})(\text{capacity, mgd})(8.34\ \text{lb/gal})}{(\text{available fluoride ion, \% expressed as a decimal})(\text{purity, \% expressed as a decimal})}$$

**Feed rate (fluoride), kg/day** =

$$\frac{(\text{dosage, mg/L})(\text{capacity, m}^3\text{/day})}{(\text{available fluoride ion, \% expressed as a decimal})(\text{purity, \% expressed as a decimal})(1{,}000)}$$

**Feed Rate (fluoride saturator), gpm** $= \dfrac{(\text{plant capacity, gpm})(\text{dosage, mg/L})}{18{,}000\ \text{mg/L}}$

**Feed Rate (fluoride saturator), Lpm** $= \dfrac{(\text{plant capacity, Lpm})(\text{dosage, mg/L})}{18{,}000\ \text{mg/L}}$

**Filter backwash rise rate, in/min** $= \dfrac{(\text{backwash rate, gpm/ft}^2)(12\ \text{in/ft})}{7.48\ \text{gal/ft}^3}$

**Filter backwash rise rate, cm/min** $= \dfrac{\text{water rise, cm}}{\text{time, min}}$

**Filter drop test velocity, ft/min** $= \dfrac{\text{water drop, ft}}{\text{time of drop, min}}$

**Filter drop test velocity, m/min** $= \dfrac{\text{water drop, m}}{\text{time of drop, min}}$

**Filter loading rate, gpm/ft²** $= \dfrac{\text{flow, gpm}}{\text{filter area, ft}^2}$

**Filter loading rate, L/sec/m²** $= \dfrac{\text{flow, L/s}}{\text{filter area, m}^2}$

**Filter yield, lb/hr/ft²** $= \dfrac{\text{(solids loading, lb/day)(recovery, \% expressed as a decimal)}}{\text{(filter operation, hr/day)(area, ft}^2\text{)}}$

**Filter yield, kg/hr/m²** $= \dfrac{\text{(solids concentration, \% expressed as a decimal)(sludge feed rate, L/hr)(10)}}{\text{(surface area of filter, m}^2\text{)}}$

**Flow rate, ft³/s** $= \text{(area, ft}^2\text{)(velocity, ft/s)}$

**Flow rate, m³/s** $= \text{(area, m}^2\text{)(velocity, m/s)}$

**Force, lb** $= \text{(pressure, psi)(area, in}^2\text{)}$

**Force, newtons** $= \text{(pressure, pascals)(area, m}^2\text{)}$

**Hardness, as mg CaCO₃/L** $= \dfrac{\text{(titrant volume, mL)(1,000)}}{\text{sample volume, mL}}$ *Only when the titration factor is 1.00 of EDTA*

**Horsepower, brake, hp** $= \dfrac{\text{(flow, gpm)(head, ft)}}{\text{(3,960)(pump efficiency, \% expressed as a decimal)}}$

**Horsepower, brake, kW** $= \dfrac{\text{(9.8)(flow, m}^3\text{/s)(head, m)}}{\text{(pump efficiency, \% expressed as a decimal)}}$

**Horsepower, motor, hp** =

$$\frac{\text{(flow, gpm)(head, ft)}}{\text{(3,960)(pump efficiency, \% expressed as a decimal)(motor efficiency, \% expressed as a decimal)}}$$

**Horsepower, motor, kW** =

$$\frac{\text{(9.8)(Flow, m}^3\text{/s)(Head, m)}}{\text{(pump efficiency, \% expressed as a decimal)(motor efficiency, \% expressed as a decimal)}}$$

**Horsepower, water, hp** $= \dfrac{\text{(flow, gpm)(head, ft)}}{3{,}960}$

**Horsepower, water, kW** $= \text{(9.8)(flow, m}^3\text{/s)(head, m)}$

**Hydraulic loading rate, gpd/ft²** $= \dfrac{\text{total flow applied, gpd}}{\text{area, ft}^2}$

**Hydraulic loading rate, m³/day/m²** $= \dfrac{\text{total flow applied, m}^3\text{/day}}{\text{area, m}^2}$

**Hypochlorite strength, %** $= \frac{\text{chlorine required, lb}}{(\text{hypochlorite solution needed, gal})(8.34\ \text{lb/gal})} \times 100\%$

**Hypochlorite strength, %** $= \frac{(\text{chlorine required, kg})(100)}{(\text{hypochlorite solution needed, kg})}$

**Langelier saturation index** = pH – pHs

**Leakage, gpd** $= \frac{\text{volume, gal}}{\text{time, days}}$

**Leakage, Lpd** $= \frac{\text{volume, L}}{\text{time, days}}$

**Loading rate, lb/day** = (flow, mgd)(concentration, mg/L)(8.34 lb/gal)

**Loading rate, kg/day** $= \frac{(\text{volume, m}^3/\text{day})(\text{concentration, mg/L})}{1{,}000}$

**Mass, lb** = (volume, MG)(concentration, mg/L)(8.34 lb/gal)

**Mass, kg** $= \frac{(\text{volume, m}^3)(\text{concentration, mg/L})}{1{,}000}$

**Milliequivalent** = (mL)(normality)

**Molarity** $= \frac{\text{moles of solute}}{\text{liters of solution}}$

**Normality** $= \frac{\text{number of equivalent weights of solute}}{\text{liters of solution}}$

**Number of equivalent weights** $= \frac{\text{total weight}}{\text{equivalent weight}}$

**Number of moles** $= \frac{\text{total weight}}{\text{molecular weight}}$

**Power, kW** $= \frac{(\text{flow, L/sec})(\text{head, m})(9.8)}{1{,}000}$

**Reduction in flow, %** $= \frac{(\text{original flow-reduced flow})(100\%)}{\text{original flow}}$

**Removal, %** $= \frac{\text{in} - \text{out}}{\text{in}} \times 100\%$

**Slope, %** $= \frac{\text{drop or rise}}{\text{distance}} \times 100\%$

**Solids, mg/L** $= \frac{\text{(dry solids, g)(1,000,000)}}{\text{sample volume, mL}}$

**Solids concentration, mg/L** $= \frac{\text{weight, mg}}{\text{volume, L}}$

**Specific gravity** $= \frac{\text{specific weight of substance, lb/gal}}{\text{8.34 lb/gal}}$

**Specific gravity** $= \frac{\text{specific weight of substance, kg/L}}{\text{1.0, kg/L}}$

**Surface loading rate** or **surface overflow rate, gpd/ft²** $= \frac{\text{flow, gpd}}{\text{area, ft}^2}$

**Surface loading rate** or **surface overflow rate, Lpd/m²** $= \frac{\text{flow, Lpd}}{\text{area, m}^2}$

**Three normal equation** $= (C_1 \times V_1) + (C_2 \times V_2) = (C_3 \times V_3)$ ***Where $V_1 + V_2 = V_3$; C =concentration, V = volume or flow; Concentration units must match; Volume units must match***

**Threshold odor number** $= \frac{A+B}{A}$ ***Where A = volume of odor causing sample, B = volume of odor free water***

**Two normal equation** $= (C_1 \times V_1) = (C_2 \times V_2)$ ***Where C = Concentration, V = volume or flow; Concentration units must match; Volume units must match***

**Velocity, ft/s** $= \frac{\text{flow rate, ft}^3/\text{s}}{\text{area, ft}^2}$

**Velocity, ft/s** $= \frac{\text{distance, ft}}{\text{time, s}}$

**Velocity, m/s** $= \frac{\text{flow rate, m}^3/\text{s}}{\text{area, m}^2}$

**Velocity, m/s** $= \frac{\text{distance, m}}{\text{time, s}}$

**Volume of cone** $= (1/3)(0.785)(\text{diameter}^2)(\text{height})$

**Volume of cylinder** $= (0.785)(\text{diameter}^2)(\text{height})$

**Volume of rectangular tank** = (length)(width)(height)

**Water use, gpcd** $= \frac{\text{volume of water produced, gpd}}{\text{population}}$

$$\textbf{Water use, Lpcd} = \frac{\text{volume of water produced, Lpd}}{\text{population}}$$

**Watts (AC circuit)** = (volts)(amps)(power factor)

**Watts (DC circuit)** = (volts)(amps)

$$\textbf{Weir overflow rate, gpd/ft} = \frac{\text{flow, gpd}}{\text{weir length, ft}}$$

$$\textbf{Weir overflow rate, Lpd/m} = \frac{\text{flow, Lpd}}{\text{weir length, m}}$$

$$\textbf{Wire-to-water efficiency, \%} = \frac{\text{water hp}}{\text{motor hp}} \times 100\%$$

$$\textbf{Wire-to-water efficiency, \%} = \frac{\text{(flow, gpm)(total dynamic head, ft)(0.746 kW/hp)(100\%)}}{\text{(3,960)(electrical demand, kW)}}$$

## Abbreviations

**C** ........... Celsius
**cfs** .......... cubic feet per second
**cm** .......... centimeters
**DO** ......... dissolved oxygen
**EMF** ...... electromotive force
**F**............ Fahrenheit
**ft** ........... feet
**ft lb** ........ foot-pound
**g** ............ grams
**gal** .......... US gallons
**gfd**.......... US gallons flux per day
**gpcd** ....... US gallons per capita per day
**gpd**......... US gallons per day
**gpg**......... grains per US gallon
**gpm**........ US gallons per minute
**hp**........... horsepower
**hr** ........... hours
**in**............ inches
**kg**........... kilograms
**km**.......... kilometers
**kPa** ........ kilopascals
**kW**......... kilowatts
**kWh**....... kilowatt-hours
**L** ............ liters
**lb**............ pounds
**Lpcd** ...... liters per capita per day
**Lpd** ........ liters per day
**Lpm**........liters per minute
**LSI**.........Langelier Saturation Index
**m**.............meters
**MG** .........million gallons
**MGD** ......million US gallons per day
**mg/L**.......milligrams per liter
**min** .........minutes
**mL**..........milliliters
**ML**..........million liters
**MLD**.......million liters per day
**ORP**........oxidation reduction potential
**ppb** .........parts per billion
**ppm** ........parts per million
**psi** ...........pounds per square inch
**Q**.............flow
**RPM**.......revolutions per minute
**SDI** .........sludge density index
**sec**...........second
**SS**............settleable solids
**TOC** .......total organic carbon
**TSS**.........total suspended solids
**TTHM**....total trihalomethanes
**VS**...........volatile solids
**W**...........watts
**yd**............yards
**yr** ............years

## Conversion Factors

**1 acre**.............................................. = 43,560 $ft^2$
= 4,046.9 $m^2$
**1 acre foot of water** ........................ = 326,000 gal
**1 cubic foot of water** ..................... = 7.48 gal
= 62.4 lb
**1 cubic foot per second** ................ = 0.646 MGD
= 448.8 gpm
**1 cubic meter of water** ................. = 1,000 kg
= 1,000 L
= 264 gal
**1 foot** ............................................... = 0.305 m
**1 foot of water**................................ = 0.433 psi
**1 gallon (US)**.................................. = 3.785 L
= 8.34 lb of water
**1 grain per US gallon**.................... = 17.1 mg/L
**1 hectare**......................................... = 10,000 $m^2$
**1 horsepower** ................................. = 0.746 kW
= 746 W
= 33,000 ft lb/min
**1 inch**.............................................. = 2.54 cm
**1 liter per second**........................... = 0.0864 MLD
**1 meter of water** ............................ = 9.8 kPa
**1 metric ton** .................................. = 2,205 lb
= 1,000 kg
**1 mile**.............................................. = 5,280 ft
= 1.61 km
**1 million US gallons per day** ........ = 694 gpm
= 1.55 $ft^3$/sec
**1 pound** .......................................... = 0.454 kg
**1 pound per square inch**............... = 2.31 ft of water
= 6.89 kPa
**1 square meter**............................... = 1.19 $yd^2$
**1 ton** ............................................... = 2,000 lb
**1%**................................................... = 10,000 mg/L
**π or pi**............................................ = 3.14

## Alkalinity Relationships

All Alkalinity expressed as mg/L as CaCO3 • P – phenolphthalein alkalinity • T – total alkalinity

| Result of Titration | Hydroxide Alkalinity | Carbonate Alkalinity | Bicarbonate Concentration |
|---|---|---|---|
| P = 0 | 0 | 0 | T |
| P < ½T | 0 | 2P | T – 2P |
| P = ½T | 0 | 2P | 0 |
| P > ½T | 2P – T | 2(T – P) | 0 |
| P = T | T | 0 | 0 |

# Appendix B: Sample CT Tables for *Giardia* Inactivation

| CT Values for 1.5 Log Inactivation of *Giardia* Cysts by Free Chlorine | | | | | | | | | | | | | | |
|---|---|---|---|---|---|---|---|---|---|---|---|---|---|---|
| pH = 6.9 | | | | | | | | | | | | | | |
| Temp. °C | Chlorine Concentration in mg/L | | | | | | | | | | | | | |
| | 0.4 | 0.6 | 0.8 | 1.0 | 1.2 | 1.4 | 1.6 | 1.8 | 2.0 | 2.2 | 2.4 | 2.6 | 2.8 | 3.0 |
| 0.5 | 94.80 | 96.80 | 99.60 | 101.60 | 104.40 | 107.20 | 109.40 | 112.20 | 114.20 | 117.00 | 119.80 | 121.80 | 124.60 | 126.60 |
| 1.00 | 89.40 | 91.36 | 93.80 | 95.80 | 98.24 | 100.84 | 102.80 | 105.44 | 107.40 | 110.00 | 112.48 | 114.44 | 116.88 | 118.88 |
| 2.00 | 84.00 | 85.92 | 88.00 | 90.00 | 92.08 | 94.48 | 96.20 | 98.68 | 100.60 | 103.00 | 105.16 | 107.08 | 109.16 | 111.16 |
| 3.00 | 78.60 | 80.48 | 82.20 | 84.20 | 85.92 | 88.12 | 89.60 | 91.92 | 93.80 | 96.00 | 97.84 | 99.72 | 101.44 | 103.44 |
| 4.00 | 73.20 | 75.04 | 76.40 | 78.40 | 79.76 | 81.76 | 83.00 | 85.16 | 87.00 | 89.00 | 90.52 | 92.36 | 93.72 | 95.72 |
| 5.00 | 67.80 | 69.60 | 70.60 | 72.60 | 73.60 | 75.40 | 76.40 | 78.40 | 80.20 | 82.00 | 83.20 | 85.00 | 86.00 | 88.00 |
| 6.00 | 64.32 | 66.12 | 67.12 | 68.92 | 69.92 | 71.56 | 72.72 | 74.52 | 76.16 | 77.96 | 79.12 | 80.76 | 81.76 | 83.72 |
| 7.00 | 60.84 | 62.64 | 63.64 | 65.24 | 66.24 | 67.72 | 69.04 | 70.64 | 72.12 | 73.92 | 75.04 | 76.52 | 77.52 | 79.44 |
| 8.00 | 57.36 | 59.16 | 60.16 | 61.56 | 62.56 | 63.88 | 65.36 | 66.76 | 68.08 | 69.88 | 70.96 | 72.28 | 73.28 | 75.16 |
| 9.00 | 53.88 | 55.68 | 56.68 | 57.88 | 58.88 | 60.04 | 61.68 | 62.88 | 64.04 | 65.84 | 66.88 | 68.04 | 69.04 | 70.88 |
| 10.00 | 50.40 | 52.20 | 53.20 | 54.20 | 55.20 | 56.20 | 58.00 | 59.00 | 60.00 | 61.80 | 62.80 | 63.80 | 64.80 | 66.60 |
| 11.00 | 47.12 | 48.72 | 49.72 | 50.72 | 51.52 | 52.52 | 54.12 | 55.12 | 56.12 | 57.72 | 58.56 | 59.56 | 60.52 | 62.16 |
| 12.00 | 43.84 | 45.24 | 46.24 | 47.24 | 47.84 | 48.84 | 50.24 | 51.24 | 52.24 | 53.64 | 54.32 | 55.32 | 56.24 | 57.72 |
| 13.00 | 40.56 | 41.76 | 42.76 | 43.76 | 44.16 | 45.16 | 46.36 | 47.36 | 48.36 | 49.56 | 50.08 | 51.08 | 51.96 | 53.28 |
| 14.00 | 37.28 | 38.28 | 39.28 | 40.28 | 40.48 | 41.48 | 42.48 | 43.48 | 44.48 | 45.48 | 45.84 | 46.84 | 47.68 | 48.84 |
| 15.00 | 34.00 | 34.80 | 35.80 | 36.80 | 36.80 | 37.80 | 38.60 | 39.60 | 40.60 | 41.40 | 41.60 | 42.60 | 43.40 | 44.40 |
| 16.00 | 32.24 | 33.08 | 34.04 | 34.88 | 35.04 | 35.88 | 36.68 | 37.68 | 38.48 | 39.32 | 39.64 | 40.48 | 41.28 | 42.12 |
| 17.00 | 30.48 | 31.36 | 32.28 | 32.96 | 33.28 | 33.96 | 34.76 | 35.76 | 36.36 | 37.24 | 37.68 | 38.36 | 39.16 | 39.84 |
| 18.00 | 28.72 | 29.64 | 30.52 | 31.04 | 31.52 | 32.04 | 32.84 | 33.84 | 34.24 | 35.16 | 35.72 | 36.24 | 37.04 | 37.56 |
| 19.00 | 26.96 | 27.92 | 28.76 | 29.12 | 29.76 | 30.12 | 30.92 | 31.92 | 32.12 | 33.08 | 33.76 | 34.12 | 34.92 | 35.28 |
| 20.00 | 25.20 | 26.20 | 27.00 | 27.20 | 28.00 | 28.20 | 29.00 | 30.00 | 30.00 | 31.00 | 31.80 | 32.00 | 32.80 | 33.00 |
| 21.00 | 23.64 | 24.44 | 25.28 | 25.44 | 26.08 | 26.44 | 27.08 | 28.04 | 28.08 | 28.88 | 29.68 | 29.88 | 30.68 | 30.84 |
| 22.00 | 22.08 | 22.68 | 23.56 | 23.68 | 24.16 | 24.68 | 25.16 | 26.08 | 26.16 | 26.76 | 27.56 | 27.76 | 28.56 | 28.68 |
| 23.00 | 20.52 | 20.92 | 21.84 | 21.92 | 22.24 | 22.92 | 23.24 | 24.12 | 24.24 | 24.64 | 25.44 | 25.64 | 26.44 | 26.52 |
| 24.00 | 18.96 | 19.16 | 20.12 | 20.16 | 20.32 | 21.16 | 21.32 | 22.16 | 22.32 | 22.52 | 23.32 | 23.52 | 24.32 | 24.36 |
| 25.00 | 17.40 | 17.40 | 18.40 | 18.40 | 18.40 | 19.40 | 19.40 | 20.20 | 20.40 | 20.40 | 21.20 | 21.40 | 22.20 | 22.20 |

Modified from *Guidance Manual for Compliance With the Filtration and Disinfection Requirements for Public Water Systems Using Surface Water Sources.* Copyright 1990, AWWA.

| CT Values for 1.5 Log Inactivation of *Giardia* Cysts by Free Chlorine | | | | | | | | | | | | | | |
|---|---|---|---|---|---|---|---|---|---|---|---|---|---|---|
| pH = 7.6 | | | | | | | | | | | | | | |
| Temp. °C | Chlorine Concentration in mg/L | | | | | | | | | | | | | |
| | 0.4 | 0.6 | 0.8 | 1.0 | 1.2 | 1.4 | 1.6 | 1.8 | 2.0 | 2.2 | 2.4 | 2.6 | 2.8 | 3.0 |
| 0.5 | 123.00 | 124.60 | 128.00 | 132.00 | 135.40 | 138.60 | 142.60 | 145.80 | 149.00 | 154.60 | 155.40 | 158.40 | 161.60 | 164.60 |
| 1.00 | 115.64 | 117.52 | 120.68 | 124.32 | 127.48 | 130.48 | 134.08 | 137.08 | 140.08 | 144.96 | 146.20 | 149.00 | 152.00 | 154.80 |
| 2.00 | 108.28 | 110.44 | 113.36 | 116.64 | 119.56 | 122.36 | 125.56 | 128.36 | 131.16 | 135.32 | 137.00 | 139.60 | 142.40 | 145.00 |
| 3.00 | 100.92 | 103.36 | 106.04 | 108.96 | 111.64 | 114.24 | 117.04 | 119.64 | 122.24 | 125.68 | 127.80 | 130.20 | 132.80 | 135.20 |
| 4.00 | 93.56 | 96.28 | 98.72 | 101.28 | 103.72 | 106.12 | 108.52 | 110.92 | 113.32 | 116.04 | 118.60 | 120.80 | 123.20 | 125.40 |
| 5.00 | 86.20 | 89.20 | 91.40 | 93.60 | 95.80 | 98.00 | 100.00 | 102.20 | 104.40 | 106.40 | 109.40 | 111.40 | 113.60 | 115.60 |
| 6.00 | 82.04 | 84.68 | 86.84 | 88.84 | 91.00 | 93.00 | 95.00 | 97.20 | 99.16 | 101.16 | 103.96 | 105.80 | 107.96 | 109.80 |
| 7.00 | 77.88 | 80.16 | 82.28 | 84.08 | 86.20 | 88.00 | 90.00 | 92.20 | 93.92 | 95.92 | 98.52 | 100.20 | 102.32 | 104.00 |
| 8.00 | 73.72 | 75.64 | 77.72 | 79.32 | 81.40 | 83.00 | 85.00 | 87.20 | 88.68 | 90.68 | 93.08 | 94.60 | 96.68 | 98.20 |
| 9.00 | 69.56 | 71.12 | 73.16 | 74.56 | 76.60 | 78.00 | 80.00 | 82.20 | 83.44 | 85.44 | 87.64 | 89.00 | 91.04 | 92.40 |
| 10.00 | 65.40 | 66.60 | 68.60 | 69.80 | 71.80 | 73.00 | 75.00 | 77.20 | 78.20 | 80.20 | 82.20 | 83.40 | 85.40 | 86.60 |
| 11.00 | 61.04 | 62.20 | 64.04 | 65.20 | 67.04 | 68.20 | 70.00 | 72.00 | 73.00 | 74.80 | 76.80 | 77.96 | 79.76 | 80.92 |
| 12.00 | 56.68 | 57.80 | 59.48 | 60.60 | 62.28 | 63.40 | 65.00 | 66.80 | 67.80 | 69.40 | 71.40 | 72.52 | 74.12 | 75.24 |
| 13.00 | 52.32 | 53.40 | 54.92 | 56.00 | 57.52 | 58.60 | 60.00 | 61.60 | 62.60 | 64.00 | 66.00 | 67.08 | 68.48 | 69.56 |
| 14.00 | 47.96 | 49.00 | 50.36 | 51.40 | 52.76 | 53.80 | 55.00 | 56.40 | 57.40 | 58.60 | 60.60 | 61.64 | 62.84 | 63.88 |
| 15.00 | 43.60 | 44.60 | 45.80 | 46.80 | 48.00 | 49.00 | 50.00 | 51.20 | 52.20 | 53.20 | 55.20 | 56.20 | 57.20 | 58.20 |
| 16.00 | 41.32 | 42.36 | 43.52 | 44.52 | 45.68 | 46.52 | 47.52 | 48.68 | 49.68 | 50.68 | 52.32 | 53.32 | 54.32 | 55.32 |
| 17.00 | 39.04 | 40.12 | 41.24 | 42.24 | 43.36 | 44.04 | 45.04 | 46.16 | 47.16 | 48.16 | 49.44 | 50.44 | 51.44 | 52.44 |
| 18.00 | 36.76 | 37.88 | 38.96 | 39.96 | 41.04 | 41.56 | 42.56 | 43.64 | 44.64 | 45.64 | 46.56 | 47.56 | 48.56 | 49.56 |
| 19.00 | 34.48 | 35.64 | 36.68 | 37.68 | 38.72 | 39.08 | 40.08 | 41.12 | 42.12 | 43.12 | 43.68 | 44.68 | 45.68 | 46.68 |
| 20.00 | 32.20 | 33.40 | 34.40 | 35.40 | 36.40 | 36.60 | 37.60 | 38.60 | 39.60 | 40.60 | 40.80 | 41.80 | 42.80 | 43.80 |
| 21.00 | 30.12 | 31.28 | 32.12 | 33.08 | 33.92 | 34.28 | 35.08 | 36.08 | 36.92 | 37.88 | 38.08 | 39.08 | 39.88 | 40.88 |
| 22.00 | 28.04 | 29.16 | 29.84 | 30.76 | 31.44 | 31.96 | 32.56 | 33.56 | 34.24 | 35.16 | 35.36 | 36.36 | 36.96 | 37.96 |
| 23.00 | 25.96 | 27.04 | 27.56 | 28.44 | 28.96 | 29.64 | 30.04 | 31.04 | 31.56 | 32.44 | 32.64 | 33.64 | 34.04 | 35.04 |
| 24.00 | 23.88 | 24.92 | 25.28 | 26.12 | 26.48 | 27.32 | 27.52 | 28.52 | 28.88 | 29.72 | 29.92 | 30.92 | 31.12 | 32.12 |
| 25.00 | 21.80 | 22.80 | 23.00 | 23.80 | 24.00 | 25.00 | 25.00 | 26.00 | 26.20 | 27.00 | 27.20 | 28.20 | 28.20 | 29.20 |

Modified from *Guidance Manual for Compliance With the Filtration and Disinfection Requirements for Public Water Systems Using Surface Water Sources*. Copyright 1990, AWWA.

CPSIA information can be obtained
at www.ICGtesting.com
Printed in the USA
BVHW021120100523
663915BV00012B/170